Mind Metrics

Brain health, as opposed to mental health, is a neglected phenomenon, less discovered but now capturing more attention. Currently, the concepts surrounding brain health are vague and no clear definition of this term exists. Assessing brain health early in life may offer insights into impending disruptions brain function. If deteriorating or poor health of the brain is detected at a young age, therapeutic steps can be taken. The definition of brain health is "the preservation of optimal brain integrity and mental and cognitive function at a given age in the absence of overt brain diseases that affect normal brain function." Recently, cognitive age has been used as a better indicator of brain health called the "cognitive clock." This book reviews, for example, fMRI, EEG, and PET scans as well as measuring certain blood and CSF biomarker levels. A new noninvasive, cost-effective, and easily implemented tool for gauging brain health is introduced. This tool can be used in rural communities as well as in low- and middle-income countries (LMIC). This book focuses on the development of this new tool for brain health measurement, as well as its cross-validation in diseased and healthy populations.

Mind Metrics
Unlocking the Science
of Measuring Brain Health

Umema Zafar

CRC Press
Taylor & Francis Group
Boca Raton London New York

CRC Press is an imprint of the
Taylor & Francis Group, an **informa** business

First edition published 2026
by CRC Press
2385 NW Executive Center Drive, Suite 320, Boca Raton FL 33431

and by CRC Press
4 Park Square, Milton Park, Abingdon, Oxon, OX14 4RN

© 2026 Taylor & Francis Group, LLC

CRC Press is an imprint of Taylor & Francis Group, LLC

Library of Congress Cataloging-in-Publication Data
Names: Zafar, Umema author
Title: Mind metrics : unlocking the science of measuring brain health / by Umema Zafar.
Description: First edition. | Boca Raton FL : CRC Press, 2025. | Includes bibliographical references. | Identifiers: LCCN 2025004775 (print) | LCCN 2025004776 (ebook) | ISBN 9781032676586 hardback | ISBN 9781032648163 paperback | ISBN 9781032676579 ebook
Subjects: LCSH: Brain | Neurobiology | Cognitive neuroscience | Mind and body Classification: LCC QP376 .Z35 2025 (print) | LCC QP376 (ebook) | DDC 612.8/2--dc23/eng/20250421
LC record available at https://lccn.loc.gov/2025004775
LC ebook record available at https://lccn.loc.gov/2025004776

ISBN: 978-1-032-67658-6 (hbk)
ISBN: 978-1-032-64816-3 (pbk)
ISBN: 978-1-032-67657-9 (ebk)

DOI: 10.1201/9781032676579

Typeset in ITC Garamond Std
by KnowledgeWorks Global Ltd.

I dedicate this book to my parents, who first sparked my curiosity and believed in every dream I dared to share, and to my children, whose laughter, love, and endless questions remind me why I continue to dream.

Contents

Preface

In recent years, the quest to understand and optimize brain health has intensified, reflecting growing awareness of the diverse factors that influence our most complex organ. Across clinical specialties, however, varied definitions and diagnostic tools often create confusion, leading to fragmented approaches and multiple, sometimes burdensome, evaluations. These are all explained in the first four chapters. This monograph seeks to consolidate insights from four of my research endeavors to present a coherent framework for assessing, detecting, and monitoring the health of the human brain.

Beginning with a qualitative investigation into the best available scales for assessing various brain faculties, the first study underscores the necessity of streamlining diagnostic tools. By integrating expert interviews with literature-based evidence, it identifies feasible, versatile, and cost-effective instruments suited particularly for low- and middle-income countries, as described in Chapters 5 and 6. Chapter 7 addresses a timely concern—chronic traumatic encephalopathy (CTE). Although definitive diagnosis often hinges on autopsy findings, this book highlights plasma neuro-biomarkers that could enable earlier detection and intervention, potentially mitigating future debilitation.

Next, Chapter 8 delves into the development of a robust formula for measuring brain health. It demonstrates how faculties that function consistently and predictably in a healthy brain may exhibit limited predictive strength—an essential insight for refining analytical models. Finally, Chapters 9 and 10 echo these findings, reinforcing the notion that integrating brain imaging, physiological tests, and biomarker measurements is critical but still challenging.

By weaving together these investigations, this book aims to guide researchers, clinicians, and policymakers toward a unified perspective on brain health. Each chapter illustrates a distinct yet interconnected facet, building a comprehensive understanding of how best to measure and maintain our cognitive well-being. It is my hope that these combined insights will spark meaningful advancements in brain research and contribute to improved clinical practices for individuals around the globe.

Acknowledgment

Before all else, I truly thank Allah Subhanahu wa ta'ala for everything so far, especially for enabling me to achieve the goal of completing this work.

I am grateful to Dr. Syed Hamid Habib who was my Ph.D. supervisor and who helped me along every point of the way. His tireless support during my sampling and then motivation during analysis and thesis writing was a great blessing.

Special thanks to my Chairperson of Physiology Department of HBS Medical and Dental College who gave me enough relaxation to let me work efficiently on my project.

I also owe gratefulness to my parents, Mrs. Saeeda Zafar and Dr. Zafar Ul Islam, without whose tireless support and motivation I would not have been standing where I am today. Special thanks to my "Compass," who always points me in the right direction. Also, to add my three sisters for being a source of comfort, and to whom I turned to in my times of need.

Finally, a lasting tribute and love for my lovely daughters, Ayza and Inaya, who have been very understanding and patient in sharing their mother's time with her career.

About the Author

Dr. Umema Zafar is an accomplished physiologist, educator, and researcher with over a decade of experience in teaching and advancing medical sciences. Currently serving as Associate Professor of Physiology at HBS Medical College, Islamabad, she holds an MBBS, M. Phil., and a Ph.D. in Physiology from Khyber Medical University, Pakistan. Dr. Zafar has been recognized with the prestigious 2022 Research Recognition Award by the American Physiological Society and has served as Vice President of the South Asian Association of Physiologists (SAAP).

Dr. Zafar's groundbreaking work includes the development of a noninvasive tool to measure brain health and a predictive formula for height estimation based on age, gender, and hand size. Her expertise spans neurophysiology, medical education, and clinical research, with numerous publications and presentations at national and international conferences.

Beyond her professional achievements, Dr. Zafar is deeply committed to social service. She volunteered in the medicine ward at Khyber Teaching Hospital during her undergraduate years and is an active member of professional organizations, including the American Physiological Society and Pakistan Physiological Society. An advocate for continuous learning, she has recently completed a hands-on course in aesthetic medicine, excelling in minimally invasive procedures.

Dr. Zafar's work reflects her passion for improving healthcare through research, education, and innovation, making her a leader in her field.

Chapter 1

Understanding Brain Health versus Mental Health

Introduction

Greetings, brain explorers. Allow me to clarify something from the get go! Brain health and mental health are two different states of well-being. Of course, they are as alike as bread and butter in a sandwich, or I should better say more like peanut butter and jelly each contributing to the mouth-watering sandwich of your health. Mental health has to do with how you feel—the state of your mind. I'm talking about the feelings that force a tear out of your eye at a weepy movie or that excitement you get when searching for money in your pocket and discovering a $20 note. But brain health? Well, it's way bigger. Cognition is the manager of the entire process, supervising all the divisions starting with memory and learning, and ending with motor function and emotional well-being, encompassing everything in between. It does not mean, merely being happy or sad, no! In fact, it is the way our brain

DOI: 10.1201/9781032676579-1

works, enabling us to spiritedly live the very best of our lives. Well, in this chapter we won't just scratch the surface—we're going to go deeper and explore the very heart of brain health. We will talk about what it is, why it is important, and why getting your brain and mind to the state of optimum health is still important even though they are linked. By the end of the chapter, you will be getting double vision—but in a positive sense. Let's jump in!

Brain Health: Physical Foundations

Neuroplasticity is a process that determines brain actions toward stress both in acute and chronic conditions and to changes in the environment. In the course of human history, great people have illustrated the neuroplasticity feature of the brain. Think about Isaac Newton who, for example, was suspected to have a learning disability when he was young. Nonetheless, due to such challenges, Newton's brain worked brilliantly, proving that the brain can adapt to challenges and achieve tremendous achievements in certain fields, like mathematics and physics. His story, in essence, is a particularly intriguing view of the plasticity and capacity of the human brain to create new connections and links. Stress reaction or changes in stimulation levels can cause the brain to change immediately, which can include both neuroplastic changes in the short term and things such as increased awareness or enhanced memory encoding. For instance, Oprah Winfrey has come out in the public multiple times and shared how she had to endure mistreatments in childhood. Her acute stress responses helped her increase her endurance and sharpened her ability to make fast decisions which turned out to be a precious asset as she forged a career in media. Although most of them were short-term adjustments, all of them provided the premise for the company's growth and stabilization in the long run. For example, immediately after a stressor, the brain may increase the active connections within the cortex in order to come up with rapid decisions and sharpened attention. These changes are usually situation-related, although they may be extended to a certain degree, to enhance the coping ability of the given individual toward stress in certain circumstances.

Chronic stress causes long-term changes in the neuronal structure which are for long periods of time but are reversible and may result in either positive or negative effects. For instance, the great composer Ludwig van Beethoven had to endure long-standing stress resulting from the condition called tinnitus. Instead of becoming despondent, his brain found other ways to work effectively—his other senses and his imagination developed a rich music capacity which enabled him to write some of his most outstanding compositions. This shows how chronic dispositions to stress can be replaced with positive neuroplastic changes giving a positive outlook to life and favorable environments. Stress hormones such as cortisol, which get released over a prolonged period, may exert measurable harm to neuroplasticity in areas that include the hippocampus that plays pivotal roles in processes such as memory and learning. Continual stress can alter conditions within the brain by decreasing neurogenesis and even destroying synaptic connections which would further aggravate cognitive abilities and make a person more susceptible to developing mental illnesses. On the other hand, positive environmental changes are effective in increasing adaptive neuroplasticity, for instance, a friendly interaction with people or taking up new learning experiences like learning a new language make the brain resilient. Neuroplasticity also reacts to internal and external environments as well as to changes in the brain. On the internal level, the patterns of activity include feelings, cognition and affective states, and motivation, which act as endophenotypes affecting the plasticity of the brain. Positive internal experiences can change the structure of the brain in ways that may help enhance certain skills and develop adaptive thinking. Externally, well-stimulating intellectual, physiologically active, and social environments contribute to neuroplasticity to enable the brain to create, adapt, and sustain all its functions. Thus, the nature of the brain seen here demonstrates the aspects of cerebral adaptability clearly reinforcing the notion of brain health as a precursor to mental health.

The term brain health relates to the enhanced processes of the brain, both structural and functional. It covers an individual's biological and neurological functions that enable his or her brain to work effectively. Functions of brain fitness are based on structural connectivity, functionality, plasticity, basic brain operations, and behavioral characteristics. Physical coherence pertains to sound and undamaged condition of the brain tissue, various other areas, and more specifically the brain's neuronal synapses. A confirmation of this integrity involves avoiding circumstances such as exposure to increased risk of injury from activities such as cycling and accidents through lack of protective clothing such as helmets. Neural network functioning is the ability of neurons to communicate with one another and that involves neurotransmitters and impulses that range from thoughts to movement. Neuroplasticity is another important characteristic of the brain that can also be effected by various factors. It means the process through which the brain can change and remodel itself in response to new experiences, in the course of learning different types of activities and actions, and in the period following injuries of the brain. There are five essential aspects of the human brain's functionality—memory, learning, coordination, movement, and creativity—all being determined by the health of the brain. It was established that nutrition, exercise, sleep, and adequate water intake are major determinants of brain health. Nutrition helps to optimize the brain functioning, and physical activity promotes neuroplasticity of the brain. Sleep is important to the brain to reinforce the memories of the past day and to flush out toxins, whereas appropriate hydration helps cells in the brain to function well. Neurodegenerative diseases are other significant threats to the brain and therefore preventing them is another reason why much effort is needed in maintaining brains' health; that said, measures to reduce risks and promote healthy aging are important in the long run.

Nurture Your Brain Garden: Steps for a Healthier Mind

Now imagine a garden, which can be always accessed but is well concealed from one's daily routine. A "brain garden" can be imagined as a dynamic, living structure whose neurons resemble plants in a landscape. Much like real gardens require sunlight and water to flourish, this internal garden also depends on essential nutrients—akin to fertilizers or water—to foster growth and development. Just as plants require frequent watering, the brain requires stimulation constantly through learning. Whether it is a new book, challenging puzzle, or a sport resumed after a long time, all new information is a sort of gentle rain that feeds your neurons and helps create new pathways. However, as you know, any garden does not only need to grow healthily but it also has to be pruned of all the unwanted crops. Stress is another factor indicating that if you do not water your brain regularly, your cells will die as a result of being choked off by weeds. Just as with the Hanging Gardens of Babylon (Figure 1.1), it is rumored that they existed only because they were pruned.

Likewise, activities, which have an impact on reducing stress, such as meditation, deep breathing, walking in the nature, function as a way of removing such weeds so that neurons perform operations without any interference. The brain can be compared to a colorful flower garden. Rich foods

Figure 1.1   Hanging Gardens of Babylon.

Photo from https://www.flickr.com/photos/hauntedpalace/232968544 and https://www.flickr.com under CC BY 2.0, https://creativecommons. org/licenses/by/2.0/

should be regarded as fertilizers for your brain garden. Get into the habit of taking fruits, vegetables, and whole grain foods because these foods will supply your neurons with the nutrients they require. While nutrition is important, you must also drink water; dehydration could be like a drought that slows down the brain garden.

Imagine your brain as a vibrant garden, where every neuron is a delicate flower or plant contributing to its overall beauty and health. But this garden doesn't exist in isolation— it's influenced by the "weather" around it. Stress can sweep through it like a raging storm, casting dark shadows over your mental well-being. On the other hand, positive experiences and meaningful social relationships act like warm sunlight, nurturing the garden and filling it with life, happiness, and growth. You must also bear in mind that nurturing your brain garden is not a one-time process. Some days the weather will be warm and pleasant, while on other days it will be cold and rainy; some days will be favorable for plants while other days for weeds. You can successfully cultivate this with a healthy lifestyle and have a rich and growing garden in your head. So, take your hypothetical watering can and gardening gloves and start cultivating this miraculous green area inside your head!

Defining Brain Health: A Multifaceted View

Cognitive health is an integrative state that is reflective of both the integrity of brain tissue and the complexity of networks within the brain. Phenomenal understanding of brain health, for example, requires us to understand the cognitive, affective, and motor processes. According to many opinions presented in the literature, there are various concepts that pertain to human optimal brain health. Neurologically speaking, brain health is about preserving the integrity of the brain tissue, ensuring that neuronal connections are functioning effectively and the brain is functioning as well as it can. This viewpoint focuses on characteristics such as sufficiency of blood circulation, neuroplasticity, and the strength of neural connections (Figure 1.2).

Psychologically, brain health for human beings goes beyond its physiological functioning, to thinking, feeling, and being able to cope with stress. A mentally healthy person might exhibit effective thinking ability, emotional quotient, and problem-solving skills in any life task. From the viewpoint of external factors affecting the brain, they are generally associated with matters of diet, exercise, and sleep. There is a positive correlation of a proper diet, exercise, and sleep with protecting the mind from neurodegenerative diseases. Let's think of the brain as an architectural masterpiece. Just like

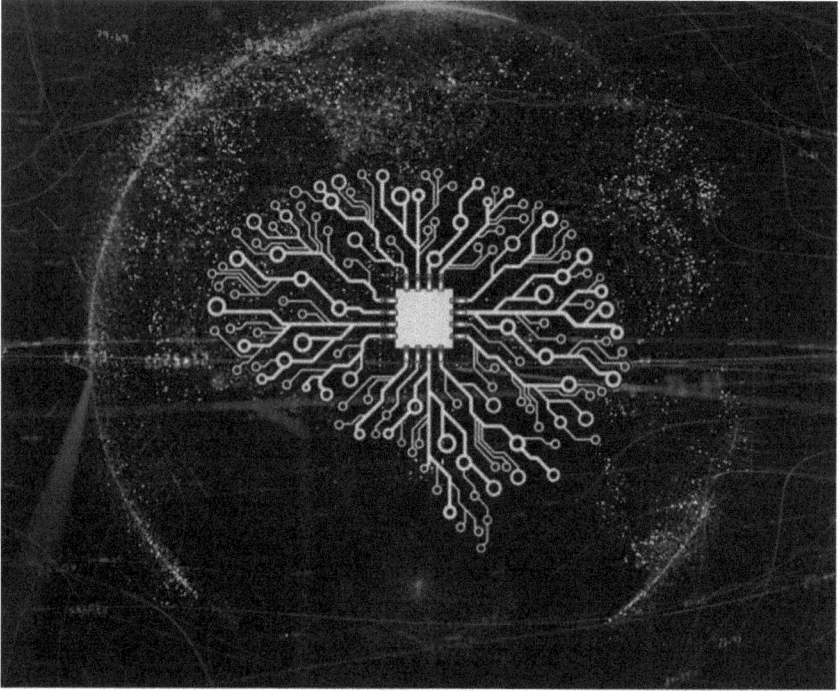

Figure 1.2   Human brain: symbolized.

*Photo from Mike MacMarketing (https://www.flickr.com/photos/
mikemacmarketing/42271822770/) and Flickr (https://www.flickr.com)
under CC BY 2.0 (https://creativecommons.org/licenses/by/2.0/).*

a skilled architect creates a strong floor plan and site design
(representing the brain's overall health), they also focus on
detailed projections, like isometric and axonometric views,
which reflect mental well-being. Even if the basic structure
of the building (the brain's neurological and functional
framework) can't be changed, revisions to the design—
through therapy, lifestyle adjustments, or interventions—can
enhance its functionality and appearance (improving mental
health). To ensure success, the foundation must be solid, and
the assets—like tools for measuring brain health—must be
carefully assessed before implementing the final design.

Navigating the Definitions: Perspectives on Brain Health

Researchers are now probing at often overlooked association between mental and physical well-being of the brain. However, at this time, there is no methodology to determine the two concepts in a more specific way. These concepts are multilayered as people from various fields, such as researchers, neuroscientists, psychologists, and experts in the field of public health, are dealing with it. The U.S. Centers for Disease Control and Prevention defines brain health as "An ability to perform all the mental processes of cognition, including the ability to learn and judge, use language, and remember."

The AHA/ASA defines optimal brain health as follows:

> The average fall off of function from the baseline performance levels that would be expected for all people of that age and in the absence of known brain or other organ system disease in terms of decline from function levels, or as adequacy to perform all activities that the individual wants to engage in.

Gorelick and his fellow researchers have described brain health as follows: "The enhancement and maintenance of brain structure and connectivity for a given age." In clinical practice, brain health is predicated in the light of neurological disorder's prevention and management. Focus should be on the early identification of abnormalities, their treatment, and functional recovery with the primary aim of improving the neuronal functioning. Public health experts can work to encourage favorable policies and conditions that allow cognitive development and improved mental health on the population level. Some of the subcategories include perception, memory, attention, and nonperceptive functions. Those who work in this area say that the brain has a remarkable degree of flexibility or neuroplasticity. Psychological views about the brain can be defined in terms of emotions and social cognition. We also need to know aspects like emotional health, coping ability, and ability to form relationships in order to understand the brain health adequately. This perspective affirms the totality of the relationship between mental and brain health as socio-emotional well-being is anchored to the highly complex neuronal networks of the brain. In other words, the conceptualizations of brain health identified in the literature are the individual tiles of a larger picture. Over time, as the studies on the brain are being done, there is an understanding that there are so many interactive components, including biological, psychological as well as social, which have bearings on the health of the brain.

Mental Health: A Debate

The WHO constitution states that speaking about health, we must understand that it implies the absence of illness and continual good physical, mental, and social conditions. There is one meaningful consequence of this definition: mental health is not merely the absence of mental disorders or disabilities. Galderisi put forward the following definition of mental health: "Mental health represents the optimum psychological well-being, which empowers an individual to make optimum utilization of his or her potentials in society in harmony with those values commonly upheld universally." According to Galderisi, mental health includes the following components: preserving one's ability to think and remembering and navigating in social environments. Cognition encompasses reasoning, solving problems, obtaining and analyzing data, storing information, and decision-making. It must be acknowledged that socialization involves the capacity to identify and regulate one's own feelings, as well as responsiveness toward other people and openness to change. Mental health is not the physical health of the mind, it is about a person being vibrant enough to understand his feelings and thoughts, handle stress, be able to work, interact with others, and be useful to the society in its broad sense. Mental health is the state of an individual's psychological, emotional, and social well-being. This touches on how one reasons, feels, and even behaves, including their ability to deal with stress, relate with others, and make decisions. Think of mental health as a six-lane highway: you've got smooth emotional roads, strong psychological foundations, minimal stress jams, fully functional driving, clear cognitive navigation, and a watchful eye on the pathological detours that might pop up along the way. Emotional health implies being able to feel and express emotions and at the same time the ability to control them. Contentment, joy, purpose, well-being, or the capacity to recover from stress also form part of mental health. Psychological state refers to an individual's thought processes, attitudes, perceptions, and the subsequent behavior. Relationship-coping is accompanied by methods of managing stress in everyday life, which is also an essential part of mental health. A person should be able to build communication, understand other people, and show empathy. It also involves cognitive utilities such as reason, judgment, focus, and perseverance of thought. Some mental illnesses, including depression, anxiety, and other psychiatric disorders, can affect these functions and therefore result in poor social relations and fluctuating emotions.

Neuroprotection's Prospects: Brain Health over Mental Health

Mental health and illness still remain taboo topics of discussion in most parts of the world. Some believe that mental health and brain health are relevant to each other, hence they should be used interchangeably. Let's consider the example of Max Cleland who was a Senator from the United States and a disabled Vietnamese War veteran (Figure 1.3). He was injured during the Vietnam War 48 years ago, and had to be discharged from military service after amputation of three of his limbs. Although he recovered from his injuries with a disability, the psychological and social debilitation remained. He received treatment for many years, which was based

Figure 1.3   The late Max Cleland in his official U.S. Senate portrait. *Photo via Wikipedia.*

predominantly on antidepressants, anxiolytics, and sleep preparations, which he soon noted were generally ineffective. Amen Clinic conducted a brain scan and the result depicted that he had post-traumatic stress disorder (PTSD) and traumatic brain injury (TBI). Clinicians then began to alter the treatment based on this new evidence and Senator Cleland saw an improvement. If the doctors had not considered the state of this patient's brain, then simply managing the symptoms of the disease would have been like shooting in the dark.

Now, metaphorically speaking, mental health can be perceived as a part or slice of brain health, to put it simply, parts of a whole. Of course, there are a lot of links between the two concepts, the physical condition of the brain is connected to the psychological condition of the human being. Brain and mental health are related to cognition, emotions, stress, and adaptation where these two aspects interact. Cognition, that is, perception, memory, attention, and ability to plan and control impulses, depends on the condition of the brain and the state of mind. When it comes to the physical architecture of the brain, the structure determines how data is processed and stored. As for the mental architecture, it determines the way this data is to be used and understood. On one hand, automatic regulation is controlled by structures in the brain, for instance, amygdala and the prefrontal cortex; on the other hand, psychological techniques facilitate the management of emotions. Stress may result in structural changes in the brain, for example, reduction of the size of hippocampus, which controls memory. This shows how the condition of the mind determines the actual state of the brain of the individual who is affected. On the other hand, a condition like traumatic brain injury may lead to mood swings or depression, thus showing the relationship between brain and mental well-being (Figure 1.4).

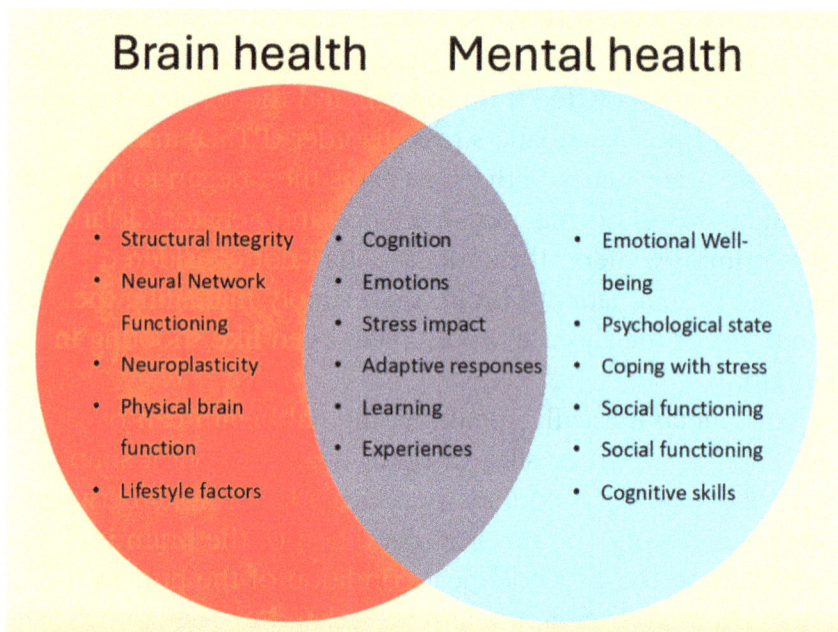

Figure 1.4   A Venn diagram showing independent and overlapping characteristics of brain and mental health.

A Holistic Concept of Health and Health Promotion

Brain health is a relatively complex concept that describes a range of structural, functional, psychological, and social determinants that affect the health of the human brain. Although mental health is concerned with the emotional and psychological well-being of the individual, it not only exists hand in hand with but is also predicated by brain health. Knowledge about the brain covers neuroscience, psychology, medicine, and public health. Living in the world today is extremely challenging and requires brain development at its finest, and indeed the foundation for a prosperous life is optimum brain health.

Brain health is more about the biomechanical processes, whereas mental health has more to do with the psychological and emotive aspects. But they are not two distinct concepts, and usually the state of one reflects the other. For example, endocrinological derangements in the neurotransmitters of the brain like serotonin or dopamine result in mental health issues. Likewise, psychological stress has been shown to have similar effects on the brain, which means that the brain also requires health-related support. Both aspects are also necessary in the holistic care. More often, treatments that combine medication and therapy are considered the most effective, because medication helps balance the brain's chemistry while therapy reshapes thought patterns—both of which ultimately improve mental health. By controlling the levels of stress and by allowing structural changes in the brain, regular mindfulness meditation can promote both better brain and mental health.

In the next chapter, we will discuss burden of brain disorders and challenges in their diagnosis. We will also discuss how and why it is a taboo topic in most parts of the world.

Bibliography

1. Liu X, et al. Cognition's dependence on functional network integrity with age is conditional on structural network integrity. Neurobiol. Aging 2023;129:195–208.
2. Lindberg S. How Does Your Environment Affect Your Mental Health? Verrywellmind, 2023. Available at https://www.google.com/url?client=internal-element-cse&cx=0903a1a109a46b6e5&q=https://www.verywellmind.com/how-your-environment-affects-your-mental-health-5093687&sa=U&ved=2ahUKEwi79tOH8IuEAxXDRvEDHVy7Az8QFnoECAAQAg&usg=AOvVaw04XH0v8ceShh0l39-q91ng (accessed February 2, 2024).
3. Wang Y, Pan Y, Li H. What is brain health and why is it important? BMJ 2020;371:m3683.

4. Cattaneo G, et al. The Barcelona brain health initiative: A cohort study to define and promote determinants of brain health. Front. Aging Neurosci. 2018;10:27–30.
5. Beudel M, Cagnan H, Little S. Adaptive brain stimulation for movement disorders. Prog. Neurol. Surg. 2018;33:230–242.
6. Sheik N. Defining the 6 key domains of cognitive function. Altoida, 2022. Available at https://altoida.com/blog/defining-the-6-key-domains-of-cognitive-function/ (accessed February 2, 2024).
7. Lahnakoski JM, et al. Synchronous brain activity across individuals underlies shared psychological perspectives. Neuroimage 2014;100:316–324.
8. Nadeem Z, Mcintosh A, Lawrie S. Mental health. WHO 2017;72:1500–1531.
9. Galderisi S, Heinz A, Kastrup M, Beezhold J, Sartorius N. A proposed new definition of mental health. Psychiatr. Pol. 2017;51:407–411.
10. WHO. Mental health. 2022. Available at https://www.who.int/data/gho/data/themes/mental-health (accessed August 9, 2022).
11. What if mental health was brain health? 2017. Available at https://www.amenclinics.com/blog/mental-health-brain-health/ (accessed Aug 9, 2022).

Chapter 2

Burden of Brain Disorders and Challenges in Their Diagnosis

Introduction

Just like your central processing unit controls and monitors all operations within a computer, your brain is the central hub for all the body's activities. Every thought you have, every movement you make, and every emotion you feel are orchestrated by an incredible conductor—the brain. But, like any complex system, it is not immune to occasional errors or failures. Sometimes, things don't work perfectly, reminding us just how delicate and intricate this amazing organ truly is. For instance, Parkinson's disease is a brain disease that shows motor complications such as shaking and muscle rigidity and therefore showing how failure at the most complex level of the human body significantly hampers one's ability to go on with life. These disruptions can present in many forms and

DOI: 10.1201/9781032676579-2

each state will have different indicators. For example, you may develop tremors, which interfere with the ability to grip objects correctly for tasks that require fine hand movements (motor deficits, such as in Parkinson's disease). The effects can be, for example, the so-called memory problems (neurodegenerative diseases, including Alzheimer's disease), or a constant feeling of sadness, present even in otherwise good weather (emotional disorders, including major depressive disorder). It is worth admitting that some people can experience hearing loss (hearing disorders like presbycusis) or cannot see clearly (vision disorders like macular degeneration). Also, due to aging-associated neurodegeneration, we may have some brain injuries that lead to dementia and early senility.

The Brain: For Dummies

First of all, let us describe some effects of specific brain-related disorders and explain what phenomena lie behind them.

Amblyopia: The Lazy Eye's Impact

"Lazy eye" is a childhood optic ailment whereby one of the eyes does not develop the normal vision and therefore will not have the sharp sight. There are several types of amblyopia: strabismic amblyopia, where the eyes are misaligned; refractive amblyopia, which occurs from a large difference in refractive errors between the two eyes; and deprivation amblyopia, which is caused by a circumstance such as cataracts that hinder vision in early years. In such cases, one eye is tasked with having to do all the work. If the vision is hampered in one eye, the brain compensates with the other eye filtering out the blurred image of the diseased eye, hence the imbalance. That is similar to having someone in a workplace with no morale—before long the poor morale is mirrored everywhere.

The area of the brain that interprets vision does not fully develop and a child might develop lifelong visual difficulties if unaddressed. It is important to identify cases such as this one early and treat them in a timely manner: for example, by taping the eye which receives the better signals. Hearing that the child has a lazy eye, it is necessary to start treatment right away, so that both the eyes work equally, and the brain sections responsible for vision are trained in equal measure.

Stroke: The Brain's Electrical Blackout

The next is stroke which is a phenomenon that looks more like an electrical blackout in the brain. Have you ever sat down to enjoy your favorite TV program and all of a sudden the lights go off? That is the state where blood flow to one part of the brain is cut off leading to the lack of oxygen supply and thus nutrients to the brain cells.

There are two main types of stroke: hemorrhagic and ischemic. A hemorrhagic stroke occurs when a blood vessel in the brain ruptures, leading to bleeding and pressure on brain tissue. On the other hand, an ischemic stroke occurs whenever a blood vessel is blocked and hence the blood supply to a certain area of the brain is stopped. This in turn leads to a series of cell death and a declined overall brain function. If the stroke occurs in some part of the brain, then simply that body part corresponding to that brain region is likely to be affected whether it is speaking, walking, or even memory. For instance, the motor stroke that harms the areas of the brain that control motor skills will cause paralysis or weakness on one side of the body. These can be weakness of one or both sides of the face, slurred speech, confusion, or loss of vision. They must be treated as early as possible to regain any use and minimize harm. A central part of recovery and avoiding additional deterioration involves rehabilitation and treatment.

Alzheimer's Disease: The Memory Heist

Alzheimer's disease is more like a stealthy memory burglar, similar to a domestic employee who occasionally steals small items. First of all, they appear to be insignificant, but with time they become more damaging, time-consuming, and inconvenient when it comes to money and the procedure of substitution. Likewise, Alzheimer's takes its toll on the individual striping them of the most valued memory and mental capabilities. It is a degenerative disease and more common in elderly patients, which causes behavioral alterations. This affects the neurons' operation, thus impounding the individuals' ability to remember something, think, or even carry out simple activities. They are all operating under that veil of gloom, where everything turns murky and shifty. There is no known treatment for this memory thief, but appropriate medications, cognitive therapies, and regular exercise help in controlling the signs and may be used to even halt the worsening of this disease.

Meniere's Disease: The Dizzy Spellcasting

Meniere's disease can feel like being on a continuous round of a merry-go-round, constantly spinning, associated with hearing loss, tinnitus, and pressure in the ear. This mental experience affects the inner ear since the brain receives conflicting signals from the balance organ and the visual system—leading to both imbalance and confusion. It is as if the brain's balance center is telling the body things that are not true. If this duality persists over a certain amount of time, the listener can develop consistent hearing loss, constant balance issues, and increased cerebral pressure, as it attempts to process these opposing details. However, this raised pressure can exert negative effects on general brain function and hence lead to fatigue, mental difficulties, and poor quality of life. Treatments for Meniere's include diet change, medication, and at times surgery.

Moderate to Severe Depression: The Emotional Black Hole

Depression is when the individual seems to scream for light in a dark hole, but even the light seems to have hard time to penetrate. Most people understand it as depression—more than just the occasional feeling of sadness, but inability to have any hope, motivation, and energy. There is an imbalance in neurotransmitters, serotonin, and dopamine accounting for mood changes. The good news is, it is possible to climb out this emotional black hole with proper treatment, medication, and support. When left untreated, depression is associated with cognitive impairment and decreased neuronal plasticity. Efforts must be focused toward treating the disease, to make the suffering individuals get back to their normal brain functioning.

Senility: Twilight of the Aging Brain

Finally, it is pertinent to look at the older generation with respect to the brain since it is apparent that the mind grows old. It is said that with old age comes wisdom and experience, but conversely, one may experience cognitive impairment, like an old computer that at times hangs or becomes slow in processing commands. Dementia is known to affect thinking ability to the extent that it hampers the normal activities of the individual. Brain-healthy practices such as mental stimulation through puzzles, socialization, and healthy eating are practical ways of preserving cognitive ability, so that the twilight years of the elderly are never dull but brimming with life.

Burden of Disorders Related to the Brain: The Burden We Carry in the Sick Bag

The worldwide burden of neurological and neurodevelopmental disorders is significant, with nearly 70% of this burden in low- and middle-income countries. This is because neurological disorders rank the highest in the burden of disability-adjusted life years (DALYs) and are the second most common cause of mortality, causing around 9 million deaths each year. While many people might overlook this problem, there is a huge burden on brain health, only 28% of low-income countries are defined and have a policy on neurological diseases unlike 64% of high-income countries.

According to estimates, the most significant sources of neurological DALYs in 2016 were ischemic heart disease (42.2%), migraines (16.3%), Alzheimer's disease (10.4%), meningoencephalitis (7.9%), and seizures (5%). By far, among the neurological disorders, Parkinson's disease is on the top, owing to the growth of elderly population (Figure 2.1).

Brain Damage/Injury

There are understood to be many types of brain injury. For example, it can be classified very broadly into acquired types of brain damage and traumatic damage to the brain. The traumatic type includes a concussion, a contusion, closed head injury, and an open head injury. The acquired type includes stroke, tumor, infectious/inflammatory processes, and hypoxic brain injury.

Yet another classification of types of brain injuries includes primary and secondary types. Primary brain injury is that which occurs at the moment or even instantly, it being a result of impact to the head. Road traffic accidents and gunshot injuries among others are some of the common examples of primary brain injury. Secondary brain injury is used to refer to changes

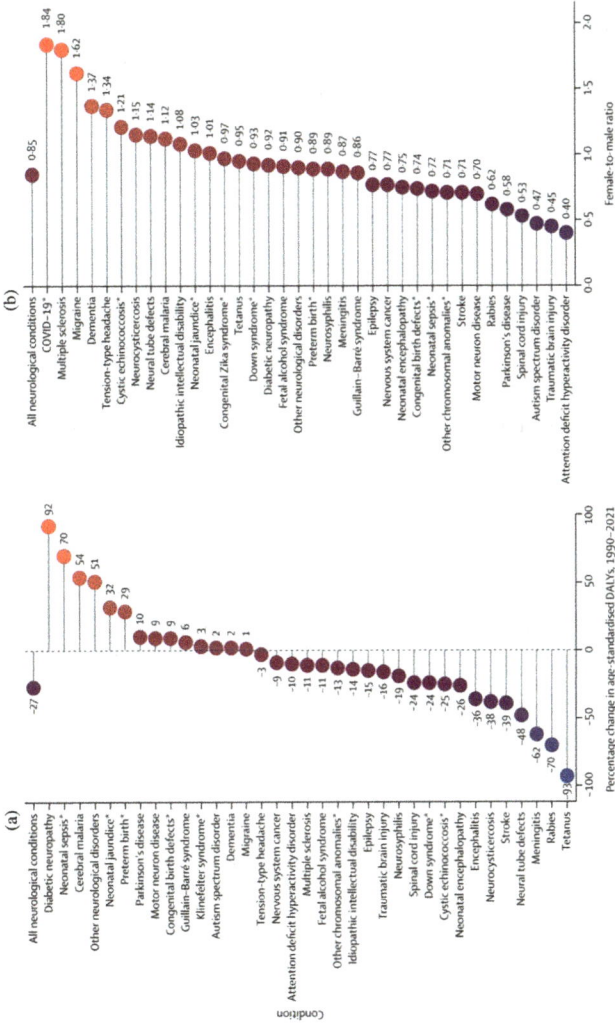

Figure 2.1   (a) Change in age-standardized DALYs, 1990–2021, and (b) female-to-male ratios in 2021 for each condition.

Steinmetz, Jaimie D. et al., Global, regional, and national burden of disorders affecting the nervous system, 1990–2021: A systematic analysis for the Global Burden of Disease Study 2021, The Lancet Neurology, 23(4), 344–381.

which occur within a brain hours to days post-injury. These involve changes that may be cellular or chemical or may involve the blood vessels and may cause some damage to the brain structure and functions, either temporarily or permanently.

Symptoms depend upon the location of the damage in the brain. Injury to the brain can be assessed anatomically (scans), physiologically (evaluations), and biochemically (biomarker tests).

Effects of Brain Injury

A head injury may have both acute (short-term) and chronic (long-term) effects. The acute effects may be related to headaches, bleeding, dislodging embolism, narrowing of the blood vessels due to thrombosis, injury of the nerves, infection, strokes, and coma states, among others. The listed conditions are progressive and can turn chronic if not treated timely. Chronic effects may relate to decline in cognitive faculties, impairments in physical activity, alterations in psychology, pathological conditions of mood, and impairment of the normal processing of sensations. Brain injuries may also predispose to pathological processes such as dementia caused by Alzheimer's disease, Parkinson's disease, and chronic traumatic encephalopathy.

Economic and Social Costs: A Ripple Effect

Imagine throwing a small stone into the still water of a pond. The first splash is the person under consideration with a brain disorder, and the ripples extend to his family, community, and society at large. These are the ripples of the disorder which each of these individuals and communities have to deal with. Let's see how these ripples contribute to social and economic challenges (Figure 2.2).

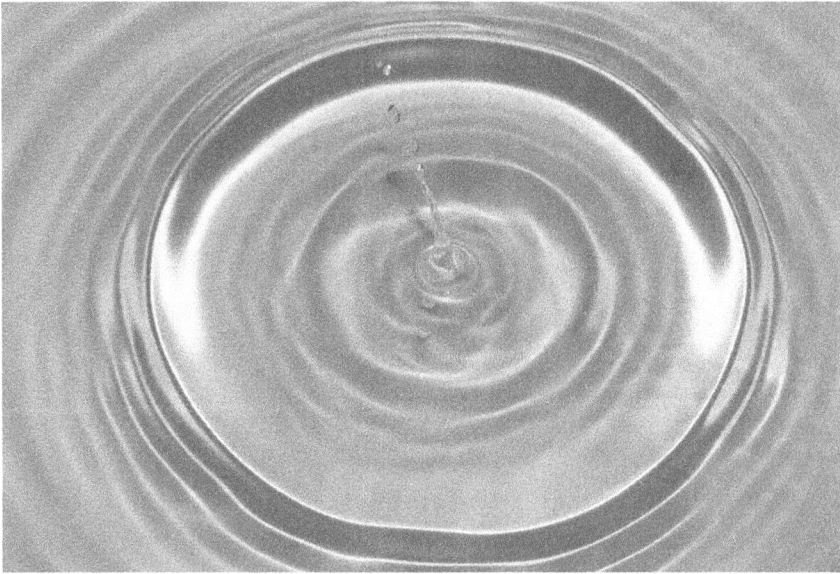

Figure 2.2   The ripple effect.

Photo from https://live.staticflickr.com/3580/3313624842_01c9fe6 e86_o_d.jpg and https://www.flickr.com under CC BY 2.0 (https:// creativecommons.org/licenses/by/2.0/).

The Individual: The Initial Impact

A person diagnosed with a brain disorder can be seen as the first ripple in a pond, triggering a series of far-reaching effects on their interpersonal world. Consider, for instance, someone who has experienced a motor stroke (hemiplegia) and now relies on a wheelchair. Their daily routine is dramatically altered due to reduced mobility and functionality, resulting in greater dependence on others. In turn, this heightened need for support often leads to an increased use of medical resources, such as more frequent doctor visits, additional medications, and specialized therapy sessions.

This impact of a disease on a person's life also has its own repercussions, and this depends on the level of the brain disorder. For example, its incidence, its morbidity, mortality, and the effects it has on the patient's functioning. Emotional well-being also becomes a growing concern as financial pressures escalate, since healthcare services can feel like a leaky bucket—no matter how much money you pour in, it always seems to slip away.

The Family: The First Ripple

The first ripple affects the family, that stress is borne by the family which often bears the emotional and financial weight of caregiving. In such type of disability, the family may need to modify their homes, surroundings, and their routine to accommodate for the accessibility and functional needs of the affected individual. Famous in this regard is the case of Stephen Hawking, who had amyotrophic lateral sclerosis (ALS) and used a wheelchair. Although his body failed him terribly, with the help of his family and caregivers, Hawking was able to enrich the realm of theoretical physics for many years. His family had to undergo multiple changes in their house setup and way of life, provide assisted living with special facilities and equipment, and many other things so that he would be able to resume his groundbreaking research and retain a certain degree of autonomy. His tale serves to illustrate both the boundless fortitude of those who suffer from neurological ailments and the toll it takes on their families. It is also the case that when a family tries to provide for a member with such conditions, the relationships become quite strained and the resources start to deplete. To illustrate, Stephen Hawking's first wife, Jane Wilde, chronicles in her book, *Travelling to Infinity: My Life with Stephen*, how his illness and hospitalization severely strained their marriage. The strain of caregiving and pursuing their dreams resulted in their temporary split.

Caregiving for long periods of time can lead to exhaustion of the caregiver, which affects the care they provide.

The Community: The Second Ripple

The second ripple touches the community. When individuals with brain disorders encounter problems with their day-to-day activities, assistance in the form of community resources or activities such as support groups, adapted transport, or community halls is usually utilized. The film "The Intouchables" (2011) by Olivier Nakache and Éric Toledano gives a beautiful example of how the whole community engages and take care of a rich quadriplegic named Philippe and his caregiver, Driss, who is from the hood. Driss is assigned to care for Philippe, who is paralyzed from neck to toes, not only to help him with physical needs but also to give him a kind of normal life he craves for. The film shows how Driss, despite having no formal training, brings a refreshing perspective and helps Philippe reconnect with the world around him. In the end, the intentions and actions of the local people in trying to assist, help, and support Philippe are quite commendable. It also shows the strain on the physical and mental attributes of the people concerned. This sort of assistance would be given to any such individual within the scope of that community reinforcing the burden and accountability on resources. And in the process, it can drain community's budgets and services as it also requires funds and people who will help create that atmosphere.

Society: The Third Ripple

Many economic and social dimensions fall within the third ripple. Everyone in a society is "affected" in this case, in both the social and economic aspects. With more and more people falling sick and having to stay in hospitals and nursing homes, healthcare costs go up. It is like a healthcare system is a kind of a sponge which is absorbing more and more resources. For example, in Japan, economic burden is a lot because of the population aging as well as the increase of senile people suffering from dementia. This places a big burden on the healthcare system as 25% of all public health spending in Japan is spent on the elderly. In addition, these losses are high in terms of productivity—brain disorder patients may not work and contribute to the economy. The financial burden extends to disability benefits and social services, placing additional stress on public finances. As the population ages, the reduced number of skilled workers can hinder economic growth, especially when many elderly individuals face cognitive difficulties.

The Unseen Costs: Social Stigma

One of the more insidious ripple effects involves the social stigma attached to psychiatric illnesses. Such misperceptions and prejudices may in turn lead to ostracism, hence complicating the process of finding help or socializing for the people suffering and their families. It is even more so the case in South Asian cultures and countries where mental and neurological illnesses are highly stigmatized. For example, the Khaled Hosseini's book, *A Thousand Splendid Suns*, features a character named Mariam who has to struggle in many aspects, including mental health and societal pressures. In this account, seeking help is not an option for her due to the cultural stigma attached to such mental illness and even talking about it is difficult for her. This particular instance goes to show how the oppressive weight of societal standards and construct can cause psychosocial problems—such as these—that are not only pervasive but also debilitative to fighting the issues proposed. This type of stigma can restrict the consumption of needed psychotropic medication, increase the level of mental disorders, and decrease quality of life.

Summarizing: A More General Approach

The entire pond reveals an interesting pattern of ripples within a ripple—each one influencing the next in surprising and intricate ways. Brain disorders have a wide range of implications economically and socially, extending over individual and collective households, communities, and society themselves. All such approaches call for more than just one strategy—namely, improvement of healthcare systems, strengthening of family systems, creation of supportive environments, and combating stigma. Understanding the ripple allows one to understand the effects that brain disorders have and the brain disorders are the motivating factors in seeking for changes that benefit not only individuals but also families and societies.

Diagnosis: Solving the Brain's Intractable Enigma

Imagine having a three-dimensional jigsaw puzzle where each of its pieces fit perfectly in multiple dimensions at the same time. Such is the case when one tries to evaluate brain-related dysfunctions, it can be equally perplexing. The process involved in the diagnosis is very complex and thus takes time and efforts. Often, you're left questioning whether a piece really belongs where you placed it. Let's now explore the depths of the dilemma called diagnosis made more complex yet interesting by comorbidities, self reporting, and overlapping symptoms.

Overlapping Symptoms: The Ambiguous Pieces

Now imagine standing in front of a puzzle and trying to find the last piece, hundreds of them are almost the same—brace yourselves for the challenge: overlapping symptoms.

Let us understand some basic things about brain disorders. Many of them have the same signs and symptoms such as memory impairment or depressive states. For example, there are significant memory problems in the course of both Alzheimer's disorder and moderate to severe depression. It is like placing a puzzle piece that has blue in it, but not quite enough to tell for sure if it is sky blue or ocean blue.

Amblyopia and stroke have similar visual problems attributed to them but with different modalities and paradigms of management. It is a tedious process of hypothesis testing. Every movement, in this case, must be spot on, lest an overzealous hand dislodges a crucial piece.

Comorbidities: The Extra Pieces

At this point, however, consider that you need to solve another puzzle: It has a few spare pieces. Comorbidities are seen as the spare pieces that complicate the diagnostic journey. Most patients suffering from brain disorders have other health issues as well. For instance, an individual diagnosed with Meniere's disease is at risk of developing anxiety because of recurrent episodes of dizziness.

Similarly, an elderly patient suffering from Alzheimer's disease may have comorbid illnesses such as diabetes or cardiovascular disease making the situation worse.

Inherent Subjectivity of Self-Reported Symptoms: The Fuzzy Parts

There are pieces that depend on how they are perceived—these are the self-reported symptoms that are subjective. The diagnosis of most brain diseases is made based on the patients' descriptions of the conditions into which many factors come into play, for example, the patients' moods, their memories, their ability to remember information, and above all the patients' readiness to search for help. It is similar to someone being asked to remember a piece of a puzzle; people's recall might be clouded by some details or even exaggerated due to recall bias. For instance, in the case of depression, patients may feel tremendous tiredness and sadness; however, the depth and impact of these symptoms are not consistent. A patient with Meniere's disease may downplay the gravity of their vertigo as they feel it is not worth reporting. The subjectivity of self-reports makes the diagnostic puzzle an even more complicated one.

Completing the Puzzle: Team Work

As piecing together an intricate puzzle requires teamwork, diagnosing a brain disorder is also a joint effort of several health practitioners, which include neurologists, psychologists, general practitioners, and even the nurses or caretakers. Everyone contributes their own piece of the puzzle, making it better able to understand the big picture.

Cognitive behavioral therapies and other approaches provide different subjective perspectives which help to address the issues. The picture becomes clearer, thanks to modern auxiliary techniques such as MRI, blood tests with biomarkers, as well as certain special tests of cognition. These devices can confirm or disprove some diagnoses, thus clearing up the tangle a bit more.

The Diagnostic Dance

The process of diagnosing different brain disorders calls for fitting together ambiguous, extra, and unclear pieces to form a coherent whole. It is time-consuming, requires expertise, and at times involves a trial and error process. The latter has allowed health practitioners to appreciate the situation and instead adopt a holistic way, which is more efficient, in diagnosing and treating such conditions.

So the next time a brain disorder crosses your mind, please think of that complex three-dimensional image. Each one, however vague, is a crucial element of the whole. Every small step toward understanding brings us closer to better brain health.

Advances in Neuroimaging and Biomarkers: The One Ring to Rule Them All

The One Ring—as depicted in "The Lord of the Rings"—is an object of immense power that reveals hidden truths but also brings great challenges (Figure 2.3). The same can be said

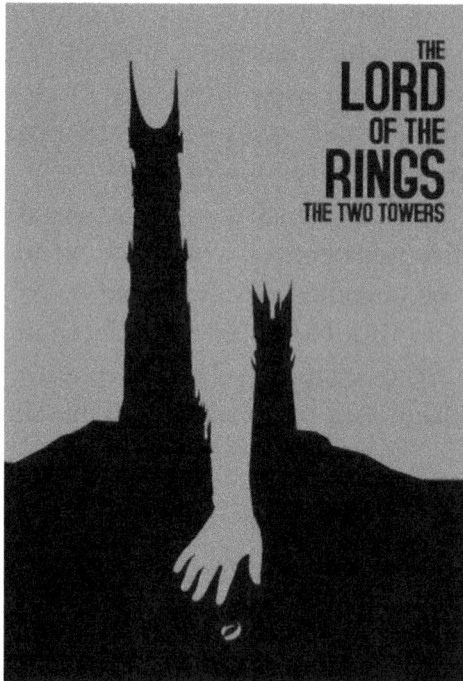

Figure 2.3   The Lord of the rings trilogy.

Image from https://www.pinterest.com/pin/203787951871925370/. Original source not specified.

about improved neuroimaging techniques and biomarkers that promise to expose particular aspects of intricate brain disorders that are hidden from view and thus allow us to see into the complex issues of the brain. However, like the Ring, it too comes with aspects that are difficult to manage if one is not cautious. Neuroimaging techniques such as an fMRI and PET scan or using biomarkers such as phosphorylated tau (pTau) gives us the power to visualize and quantify brain activity, but these also come with a challenge in their interpretation. Misrepresentation of the images could lead to misinterpretation of the results as in the case where the Ring was misused by individuals who were more ambitious than careful. Still, let us see what those new instruments are doing for gauging of brain health and why controlling their power requires as much wisdom as technical skill.

Neuroimaging along with biomarkers has transformed diagnosis and comprehension of brain disorders. Imagine wearing a powerful ring that can illuminate the hidden, allowing one to see clear pictures of what goes on inside the brain. Imaging techniques such as MRI, fMRI, PET scans, and real-time EEG monitoring enable us to visualize intricate structures and dynamic processes such as electrical activity within the brain and provide insight on the dysfunction wherever it is located.

For example, during the course of Alzheimer's disease, brain imaging can help quantify the amount of brain shrinkage and assist in the differentiation from other forms of dementia. Some markers such as neurofilament light (NfL), neuron-specific enolase (NSE), pTau, glial fibrillary acidic protein (GFAP), neurofibrillary tau (N-tau), and S100 calcium-binding protein B (S100B) can also detect neurodegenerative processes before it becomes clinically overt.

Just as the One Ring had its antagonists and obstacles, so do these innovations. For better or for worse, the more the information is given, the more hard it is to comprehend. Try to analyze what do the magical symbols inscribed on the Ring of Power mean to its bearer—it is an interesting realm, but it is also scary, just like the outcome of combining several neuroimaging tests and analyses of biomarkers. The soothing drawback of this technology is that every single scan throws out numerous alternative diagnoses. Difficulties in conclusion owing to symptom overlapping or comorbidities can make comprehending those findings even more complex. Thus, elevated levels of pTau, for instance, would be suggestive of Alzheimer's disease; however, this level may also be present in other diseases that are neurodegenerative in nature which leaves a gray area. Likewise, in cases of motor neuron disease, biomarkers such as NfL and GFAP may also be elevated signifying that there is damage being done to the neurons. Similar to these, neuroimaging may show some alterations in the structural brain components; however, their value can only be appreciated when all the factors concerning a patient's health are examined.

The One Ring is not an easy artifact to master, scientific anatomical imaging and any developed markers are no exception, they require continuous improvement. Researchers and clinicians must keep up with the latest advancements and improve their techniques over time.

> *Neuroimaging:* With the progress of imaging techniques, pictures of the brain become clearer and more defined. However, it also means that we have to get better at understanding those pictures. New programs and approaches are continuously developed to assist us in making conclusions out of the huge volumes of this information obtained from scans—it is like getting an upgraded version of the power of the Ring.
>
> *Biomarkers:* The advent of new biomarkers is of great significance; however, it entails that we need to know

more about how they work and their mechanisms. It is like making a new ring that has sharper inscriptions and it proficiently serves its purpose.

The burden of the use of this potent instrument is that it requires great skill. Indeed, it does not take away the fact that the health providers must be trained on not only using the devices but also appreciating their shortcomings. Taking it to the other extreme or position is not the solution.

The Influence of Heredity: Imagine a LEGO® House Missing a Few Key Bricks

Picture building a house out of LEGO® bricks, where each brick represents a gene. Much like the design of a LEGO® creation depends on how the pieces fit together, the architecture of the brain relies on the interplay of its genetic components. Some of these "bricks" come from your biological makeup, while others reflect environmental factors. When certain vital bricks go missing, it can weaken the structure and lead to challenges in brain health.

Geneticists know they can't simply ignore these missing pieces, as they play a pivotal role in determining a person's risk for various brain-related conditions. In essence, assembling a healthy brain resembles collecting all the right bricks—both unique and ordinary—to complete a well-crafted design. A good example is that of familial Alzheimer's disease, and how specific mutations predispose individuals to the illness. There is a greater risk of acquiring illness where there is a family history of the condition. Huntington's disease is another instance, a disorder that is caused by the alteration of one's HD gene which is referred to as the Huntingtin gene. Transmission of the disease follows an autosomal dominant inheritance, which means that there is

50% chance of inheriting the gene from the affected parent. Most recently, famed American folk singer Woody Guthrie suffered from this condition which highlighted the plight of many who suffer from this degenerative ailment.

Genetic Risk Factors: The Unnoticed Components of the Picture

Now, reflect on the unseen factors that enhance the possibility of suffering from a brain disorder. These genetic risk factors resemble the extra bricks tucked away in the LEGO® box that one buys. Certain genetic alterations can predispose an individual to certain illnesses, like depression or anxiety. These predispositions do not mean that the individual will develop a disorder; however, they do increase the chances of such a development, particularly when other aggravating circumstances such as the environment are present. It is like having a weak foundation that makes the house more vulnerable during storms. Known examples are the genes BRCA1 and BRCA2, whose presence significantly predisposes an individual to breast and ovarian cancers. One such case is where actress Angelina Jolie underwent preventive surgery after learning of her carrying the disease-associated BRCA1 mutation, lowering her chances of developing breast cancer. Such a situation is akin to making alterations to a LEGO® house in order to make it stronger against potential threats.

Interplay of Multiple Conditions

Some groups of blocks within a LEGO® house are interlinked resulting in the presence of comorbidities where two or more conditions are simultaneously observed (Figure 2.4). For example, consider the genetic abnormalities like mutations in the APOE gene which is primarily responsible for the development of Alzheimer's. Such genetic problems may also

Figure 2.4   Harry Potter–themed LEGO® set.

Image from https://ie.pinterest.com/pin/721420434036513259/.
Original source not specified.

increase the risk of developing certain cardiovascular diseases, for instance, hypertension, atherosclerosis, and heart failure. Likewise, the symptoms of metabolic syndrome can aggravate in a patient who has depression, as physiological alterations in the metabolic system as observed in nutrition-related diseases like obesity, insulin resistance, chronic inflammatory processes, in turn, worsen depression and therapeutic compliance is difficult.

An illustration of this scenario is Winston Churchill, who had bouts of depression which he referred to as his "black dog," he also had issues with obesity, thus complicating his fight against depression. This interrelatedness makes diagnosis and treatment tricky, as addressing one condition could impact another. It is like trying to adjust one brick without disrupting the entire structure.

The Conundrum of Genetic Counselling

Genetic counseling is akin to employing the services of an experienced architect possessing all the nitty-gritties of architecture and can guide you through the complexities of building houses. Genetic counseling involves assessing the chances of developing any hereditary disorders, the provision of relevant information about prenatal diagnosis, and aiding the families in making informed decisions.

Evaluation: Just like how an architect will check a building for any weaknesses, a genetic counselor will look at the risk factors present in the family history of an individual.

Genetic Testing: This is like the process of checking the building materials for the cause of the specific problem or anomalies that can be pointed out and generalization made about the building.

Management: Counselors assist patients in testing, prevention, and management such that it is similar to where an architect advises on general structural enhancement or what materials should be utilized in constructing a strong building.

The Influencers of Brain Disorders

Amblyopia

Genetics play a role in the embryological development of the eye which puts certain children at risk to this condition. Such children are those who have an incidence of strabismus in the family. Cervical strain also increases the risk of amblyopia owing to abnormal development of the visual system in the course of growth, especially in premature babies.

Stroke

The risk of stroke is quite similar to that of constructing a house with a faulty base. For instance, stroke is more likely among individuals presenting with familial tendencies for hypertension and high cholesterol. Furthermore, some genetic causes for the advancement of such thrombotic conditions also exist, such as mutations in certain clotting factors; therefore, these aspects need management in terms of lifestyle and health factors in addition to pharmacological interventions.

Alzheimer's Disease

Alzheimer's risk is greatly increased by specific alleles such as the APOE-e4, which acts more like a hidden Achilles' heel that gradually reduces the affected individual's ability to cope with the disease.

Meniere's Disease

Genetic factors can also influence the susceptibility of some individuals to ear-related problems that are characterized by changes in balance and hearing. For instance, the COCH and FAM136A genes have been associated with Meniere's disease, which explains why it is considered to be hereditary.

Depression Moderate to Severe

Environmental factors combined with genetic risk factors such as SLC6A4 and BDNF gene variation imbalances are causative in changes in serotonin, dopamine, and norepinephrine levels as well as cortisol, making people prone to depressive episodes.

Aging Population

Age brings out inherent genetic defenses that influence factors such as memory, cognition, and general brain well-being. Biomarkers that are associated with brain aging most frequently include amyloid beta, tau proteins, and downstream inflammatory markers like C-reactive protein (CRP), which turn out to be pathological in most as age progresses causing cognitive dysfunction and increasing the risk of cerebral degenerative diseases. For instance, Alzheimer's disease is characterized by the presence of amyloid beta plaques that interfere with neuronal signaling due to the deposition of beta-amyloid peptides. Disruptions in tau proteins usually result in the formation of neurofibrillary tangles that make neurons dysfunctional. High CRP levels are engrained with chronic inflammatory processes that destroy cognitive functions and increase the chances of dementia.

Building Better Foundations for Homes

Achieving the objective of erecting a stronger house also requires comprehending these weaknesses and the significance that follows suit. One risk management option is genetic counseling and regular physical checkups harvesting a healthy diet in an effort to boost the brain's defense mechanisms. In a similar fashion to ensuring that a building is in perfect condition, seeking help and making the right choices can aid in developing an improved way of taking care of the brain.

The Impact of Lifestyle Choices on the High Performance of the Brain: Daily Brain Cuisine

Just like the body, the brain thrives on a daily intake of nourishing ingredients. Some nutrients assist in the development of the brain, while others have the opposite effect. Furthermore, let us analyze the role of sleep, nutrition, exercise, and management of stress within the ambit of treatment of brain disorders, thereby turning the diet of the brain into a feast.

Sleep is, throughout, the best form of cure for the active and working brain rather like a thick warm soup that is restorative and comforting. Sleep plays an important role in memory formation, detoxification, and getting the brain ready for a new day. Missing in on sleep however is comparable to overcooking food where the pot has been left on the stove for too long, and so the contents have burnt and the taste has been lost.

Proper nutrition and a well-balanced diet are imperative in ensuring optimal brain health. A dietary pattern that consists mainly of highly processed and sugary refined carbohydrates and unhealthy fats is like offering your brain something unhealthy and greasy, it clogs the system and slows everything down. In Meniere's disorder, symptoms can be mitigated by restricting salt intake, while for an individual with depression, tryptophan-containing foods can be helpful as it increases serotonin levels.

Exercise helps to strengthen the brain, it is a brain's physical instructor. Physical exercises increase the blood supply of the brain and promote the development of new neurons which improve an individual's cognitive performance. It is like workout where the brain is exercised on a constant basis so that it focuses and remains alert. Jane Fonda, an actress, fitness guru, and activist known for her dedication to promoting health and wellness, believed that, exercise is a great way to prevent the aging process and maintain not only a healthy body but also a healthy mind.

Engaging in activities to relieve stress is akin to undergoing a detox treatment for one's mind by erasing all digital footprints to make one's brain work effectively. Chronic stress is an over-spiced cuisine: If stress were a dish, and the cook were to add spices but did not limit them. High-stress limits activate cortisol that eventually affects human brain cells, which negatively affects one's thinking abilities.

If one experiences anxiety or depression, it is vital for one to cope with those stressors. These include practices like mindfulness and meditation, which are more about focusing on your breathing or any hobby that takes your attention away from any stress-inducing factors. Creating an environment with minimal stress is like arranging your kitchen—it facilitates everything.

The Daily Diet of the Brain: A Balanced Dharma

To enhance the functionality of the brain, you must have a well-balanced life. Here is a possible menu on what the brain's meal on a daily basis should feature:

Sleep: Sleep at least 7–9 hours of sound sleep on a daily basis. Implement a relaxing pre-sleep regime.

Nutrition: Refer primarily to a diet of unrefined carbohydrates, whole fruits and vegetables, lean protein sources, and healthful oils. Limit added sugars and processed products.

Exercise: A minimum of 30 minutes of moderate exercising should be done daily. Take part in aerobics or other exercises that you like so that you do not give up.

Stress Management: Regularly engage in activities that help manage stress. Learn which of the exercises works best for you such as yoga, meditation, or doing something creative.

The Unsung Fear: Reflection and Understanding

Let's imagine a street on a sunny afternoon. The sun casts clear shadows, but these shadows can also alter and change the perception of the original shapes of the objects causing fear and illusions in the audience. The stigma surrounding brain disorders equally does that by making people's perception of the reality of these diseases worse than how they actually are, thus creating fear and lack of understanding. What comes next is how social campaigns could do away with these complexes by appropriately educating the society.

For example, if someone mentions the word Alzheimer's or depression, people often get agitated and afraid because of the twisted views cultivated and held by the society. It is like looking at a benign tree only to see the exaggerated shadow of the monster. Lazy eye might not seem serious, but the shadow of ignorance makes people overlook how it can significantly affect a child's growth. There is a lot of fear surrounding stroke and the shadow cast is that it is not a treatable condition but of one that ends in death. Then there is Alzheimer's which carries an air of inevitability, making people think it is an unstoppable decline rather than a condition where early intervention can make a significant difference. The condition known as leprosy was regarded with a high level of discrimination in the time of Jesus, despite being treatable. Leprosy was shunned, and patients were often confined, which only added to the hysteria surrounding the disease rather than utilizing the treatment methods.

Social awareness is like a torch light, coming in to do away with every shadow and putting everything that concerns mental illness into the proper perspective. It is possible to change attitudes toward society's understanding so that there is less fear and more understanding and compassion instead of accurate information. Just as a single light can illuminate a whole room, public education on brain disorders can have a cascading effect, destigmatizing and encouraging people. The awareness that Alzheimer's disease is more than just "getting old," but rather, a well-defined sickness of the brain, or the possibility of recovering from a stroke after treatment, all these elements help to reduce the stigma.

New therapies and increased focus on patient education are allowing us to explore new frontiers—all for the betterment of the lives of individuals with brain illnesses.

Picture standing on the deck of a ship in the early morning, admiring the first slivers of light above the horizon. This captures the significance of studies on brain health that are still in their infancy—the emergence of breakthroughs that will enrich and augment the scope of research.

Better insight of the pathophysiology of Alzheimer's is indicating that it is possible to realize early Alzheimer's diagnosis and treatment strategies. New drugs which address amyloid pathology and tau pathology are at the forefront in managing the disease and its effects on the mind. For example, monoclonal antibodies such as aducanumab are engineered to bind to and clear amyloid-beta plaques concentration in the brain to slow cognitive deterioration. Also, anti-tau therapies, including tau aggregation inhibitors, are being developed to combat neurofibrillary tangles to safeguard neuron health and protect brain function.

Newer cerebral imaging techniques and clot-retrieval methods are enhancing the prognosis of primary stroke patients. Equipment used in rehabilitation, including robots and virtual reality, is facilitating recovery and assisting patients in enhancing their autonomy and living standards.

Emerging treatment options such as perceptual learning and video game therapy are helping in vision advancement in children suffering from amblyopia. These conducive treatments are changing the way this prevalent childhood affliction is managed. In contrast to the traditional eye patching therapy which entails allowing the weaker eye to work under duress by covering the stronger eye, perceptual learning and video game therapy are more interactive, enjoyable, and entertaining. Perceptual learning entails training of the eyes to enable them to process the information, whereas in video game therapy, the child plays games that aim at the underperforming eye. These newer techniques are unlike patching, which can be boring and distressing for children. They offer a more playful and enjoyable modality of enhancing vision.

The investigation into the possible causes of Meniere's disease has resulted in better management of the disease. New horizons opened by the discoveries in gene therapy and personalized medicine will help in developing specific therapies that will help to alleviate the symptoms and enhance the quality of life.

Factors such as improvement in the understanding of brain chemistry and neural pathways have led to development of new antidepressant drugs in combination with physical treatment such as transcranial magnetic stimulation (TMS), ketamine therapy, cognitive behavioral therapy (CBT), and neurofeedback therapy. TMS is where nerve cells are activated by means of magnetic fields, ketamine therapy is an infusion therapy targeting the treatment of persistent depression, while CBT focuses on understanding and altering negative ideas of patients. Neurofeedback therapy is a depression treatment technique that integrates self-regulation through visual information about the brain wave activity based on real-time feedback. It has shown to be effective in treating depression. These therapies offer hope for human beings who have not responded well to traditional treatments.

Research on healthy aging is uncovering ways to maintain cognitive function and prevent dementia. Interventions focused on lifestyle changes, social engagement, and cognitive exercises are helping older adults stay active and sharp.

With emerging research, innovative therapies, and unwavering advocacy, we are heading toward a future where brain disorders are understood, managed, and, ultimately, overcome.

Bibliography

1. Wang Y, Pan Y, Li H. What is brain health and why is it important? BMJ 2020;371(9):m3683. doi: 10.1136/bmj.m3683.
2. Cattaneo G, Bartrés-Faz D, Morris TP, Sánchez JS, Macià D, Tarrero C, et al. The Barcelona brain health initiative: A cohort study to define and promote determinants of brain health. Front. Aging Neurosci. 2018; 10(7354):27–30. doi: 10.1038/475027a.
3. Nadeem Z, Mcintosh A, Lawrie S. Mental Health (Vol. 72, pp. 1500–1531). WHO, 2017. Available at https://www.who.int/health-topics/mental-health#tab=tab_2.
4. Galderisi S, Heinz A, Kastrup M, Beezhold J, Sartorius N. A proposed new definition of mental health. Psychiatr. Pol. 2017;51(3):407–411. doi: 10.12740/PP/74145.
5. Mental health. WHO. 2022. Available at https://www.who.int/data/gho/data/themes/mental-health (accessed Aug 9, 2022).
6. What if mental health was brain health? 2017. Available at https://www.amenclinics.com/blog/mental-health-brain-health/ (accessed Aug 9, 2022).

Chapter 3

Factors Affecting Brain Health

Introduction

The brain is an amazing machine. In contrast to its portrayal in science fiction literature, this extraordinary biological wonder located inside our skulls takes care of various functions as complex as cognition and audition. The circadian rhythm and the daily dance of the sunrise and sunset match in a seamless manner. It is almost as if these events orchestrate in a near-perfect fashion, setting right both themselves and everything that orbits them. Likewise, one more instance would be maintaining the internal environment of the human body healthy, free from diseases and thus living a harmonious life.

The brain's roles can be broadly grouped into four main types: perception, action, cognition, and emotion. In other words, Brodmann divided brain into 52 pieces, each with its unique anatomy and set of functions (Table 3.1 and

DOI: 10.1201/9781032676579-3

Table 3.1   Brodmann's Areas Corresponding to Brain Functions

S. No.	Functions	Brodmann's Areas
1	Motor	4, 6, 26
2	Cognition	9, 10, 21, 29, 31, 34, 35, 36, 39, 46, 52
3	Vision	8, 11, 12, 17, 18, 19, 20
4	Emotion	23, 24, 25, 28, 29, 30, 31, 33
5	Hearing	22, 41, 42
6	Cognition and emotion	32, 48
7	Cognition and vision	37, 38
8	Hearing and emotions	40
9	Vision and motor	13, 14, 15, 16
10	Taste, smell, language, and somatic senses	5, 7, 27, 43, 44, 45, 47

Figure 3.1). In order to assess how healthy your mind is holistically, you must consider all or most of those aspects listed here. Nine out of ten human brain faculties are more than enough to represent different functional specializations. Consequently, we skipped taste, smell, language, and somatic sensations. Attempting to measure brain health by trimming out domains associated with these functions would have made data collection processes tiresome and intricate. The nine example functions have an absolute correspondence with each other. Since cognition appears thrice, motor function twice, vision twice, hearing twice, and emotional well-being thrice, it is only reasonable to merge multiple functions into one. By incorporating these areas, the domains under discussion will reduce to cognition, motor function, vision, audition, and emotional well-being.

Quantitative measurement of different functions of the brain corresponding to the different areas of the brain can

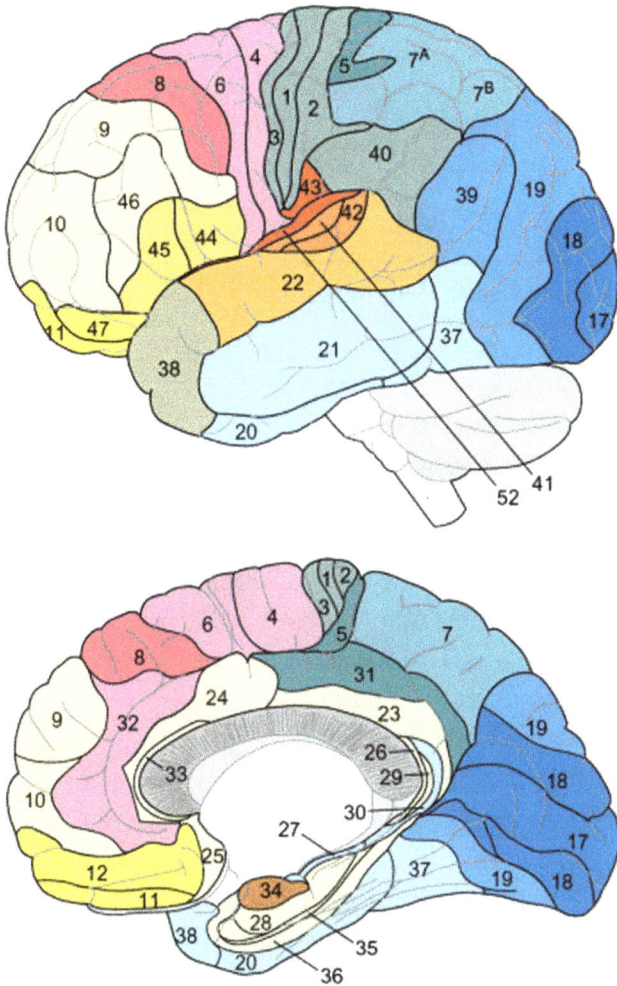

Figure 3.1   Brodmann's areas.

Image from https://www.sciencedirect.com/topics/neuroscience/brodmann-area-8

be made. For instance, one can quantify motor function corresponding to the motor cortex, cerebellum, and basal ganglia by checking on hand dexterity, balance, and strength/power. If a person has normal cognitive abilities, this implies that his/her brain, particularly the prefrontal cortex (PFC), is in good health (Figure 3.2).

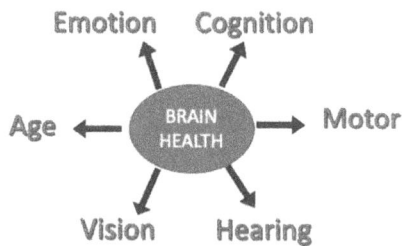

Figure 3.2   Brain health determinants.

Cognition

Cognition: The Ultimate Supercomputer

Sherlock Holmes had an interesting way of handling problems in life. He emphasized: "My mind rebels at stagnation. Give me problems, give me work, give me the most abstruse cryptogram or the most intricate analysis, and I am in my own proper atmosphere." To avoid laziness, he asked for the hardest problems that required deep mathematical knowledge like those found within abstract mathematics itself. In the same way, our minds work like top detectives who are constantly analyzing different evidences. The more it is used, the more it develops; the more it is put to rest, the lower its capability gets. Cognition involves just a few main tasks and these are as follows:

Perception of Data: Your brain is like a radar system on an air traffic control center just as the telephone receiver; hence it receives signals from sensory organs such as eyes, ears, nose, or else those found on your muscles which aid motion.

Understanding/Interpretation of Data: Data once it reaches the brain will undergo serious scrutiny; as the brain sifts through for relevant clues that will enable you to process only what is necessary while discarding those that are not important or do not concern you. At the beginning, before any response can be given, this could be referred to as sentinel relay areas responsible for reception, sorting, and delivery to other parts of our brain for further processing.

Remembrance/Recall of Information: Your memory stores everything important about your plans, whether long-term or short-term as well as for operations in progress. In analyzing the current situations these days, one can remember this stuff, thus using previously acquired knowledge plus being aware of the environment around him/her, both spatially and temporally through recovery of memories rather than storing them. We can recall them whenever needed, including permanently or temporarily storing them whenever it is necessary.

Reaction: When all this "receiving, interpreting remembering" business is done—now is the time for action! It makes decisions on how best to respond as well as ensures its perfect execution without any problems whatsoever either through moving muscles or regulating hormones through modifying processing pathways upward or downward.

But How Does It Really Work?

Some pieces of data, or clues, are being received via sensory organs. The visual information is captured by your eyes, sounds are heard through ears, the sense of smell comes from the nose, and nasopharynx guides you what thing is harmful through its scent receptors (chemical receptors).

Once the sensory data is in, it's up to your working memory to keep things running smoothly. This is where the magic happens—your brain starts understanding and interpreting the data. It is like having a super-smart assistant who organizes everything for you so that you can focus on the big picture.

Your sensory organs are like army officers who take charge from the front line during battle periods. Your eyes see images clearly while ear canal ossicles vibrate when sound waves strike them, these signals then reach brain cells where they are transformed into recognizable signals, the nose detects scents, and proprioceptors help you navigate the physical world. Together, they send a constant stream of data to your brain. Now, the brain's working memory is like a team of hyper-efficient librarians. They not only catalog and store information but also make sure it is easily retrievable when you need it. It can be considered the ultimate library where everything is perfectly organized and instantly accessible.

Once the brain has processed the information, it is showtime! The brain decides on the best course of action and executes it. Whether it is answering a tricky question, dodging an obstacle, or simply deciding what to have for lunch, your brain handles it all with finesse. So, cognition is your brain's way of being the ultimate problem-solver, memory keeper, and action taker, all rolled into one amazing supercomputer (Figure 3.3).

Central Processes in Thinking

The act of being unable to remember where our car keys are left in order to unravel a complex riddle originates from short-term memory which operates like a supercomputer. Its functionality has been likened to that of a mental notepad that temporarily stores and processes data as in a computer's random access memory (RAM), hence enabling us to think through problems in real time. The best way is to picture your working memory functioning just like having three leading actors—verbal working memory (VWM), visual spatial working memory (VSM), and attentional control system (ACS).

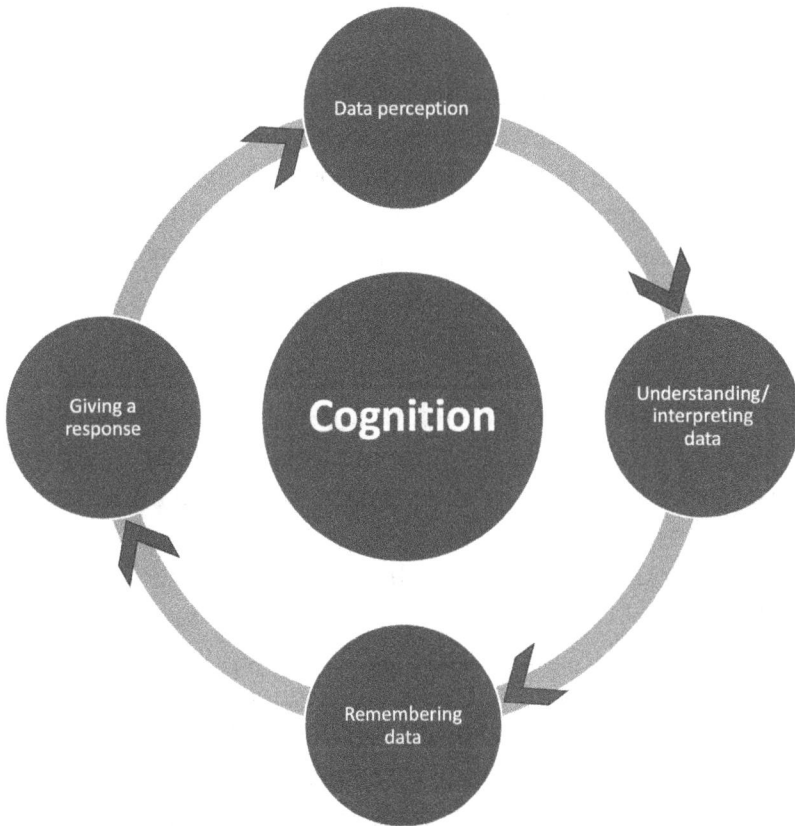

Figure 3.3   The cognitive process.

An artist who draws scenes from memory without looking at them directly is relying on his/her visual–spatial skills. Now let us take it from this point: Picturing back where your favorite café would be is an illustration of visual–spatial working memory in operation. The ability of both prefrontal and occipital cortices is engaged for this feat.

Likewise, think about that song which lit your mood to happiness or the latest gossip you just heard. This is the role for verbal working memory where words are processed right in the head. It sounds more like an intelligent personal assistant perched within the PFC, who deals with the cortical processing and creates ideas that flow smoothly outward into speech.

There are some individuals who can remain focused on what they are doing even when they are surrounded by a lot of noise. Have you ever tried shooing away a housefly while concentrating on something else entirely? This attention control system is responsible for maintaining focus despite unnecessary interruptions resulting from other thoughts or events. It helps to only hold on to three or five things at once in order to prevent valuable stuff from falling through the cracks while pulling highly relevant details during sudden decision-making moments, dumping the unfamiliar fragmented ideas. Under emergency circumstances, it can be a life-saving attribute.

If your brain is as fast as high-speed internet services, then you have good brain processing pace. Quick thinkers usually have higher IQs owing to their lightning response times. There has always been a debate as to whether one should have general expertise or specialize in certain aspects; all-rounded people say, "jacks-of all trades but master of none. They achieve generality only at the price of broad ineptitude." "Success" is a relative term with regard to the goals set, motivation behind it, and rate of achievement. However, generalizing the above debate, intelligence is high if executive function is optimum. If the ability to think and plan (executive function) are hindered, then the performance of the individual and, in particular, task goes down. Mental stimulation such as learning a new language, solving a puzzle, or playing a strategic video game improves brain health by synapse-building and pruning.

The brain thrives on challenge! For example, synaptic-building or pruning in response to mental stimulation like learning new languages, solving puzzles, or even playing strategic video games can boost your brain health. Hence, plasticity helps keep your brain young and vibrant. In structural plasticity, there is dendritic shape modifications, while functional plasticity allows other brain regions to perform functions of those brain areas that are no longer functional. It is like having a team of champions in your head! Although the mechanisms behind neuroplasticity are intricate, some research suggests the involvement of *N*-methyl-D-aspartate (NMDA) receptors. Functional magnetic resonance imaging (fMRI) and diffusion tensor imaging (DTI) have been used to research circuit modification, suggestive of plastic changes in the brain.

Diseases Associated with Cognitive Impairment

Cognitive impairment is a syndrome that refers to a decline in one or more brain domains. It encompasses diverse conditions, with the most prevalent being Alzheimer's disease, dementia caused by Parkinson's disease, and traumatic brain injury, as well as stroke among others. The symptoms may vary depending on the type of disorder, but generally they include amnesia that deals with memory loss, behavior disorders mostly referred to as mood changes, i.e., depression, and mania; impulse control disorders include loss of judgment while thinking (executive function), and processing speed deficits. The biggest risk factor for cognitive impairment is age; this condition affects nearly 5.1 million Americans aged 65 years or older who have Alzheimer's disease. Although many believe that Alzheimer's is predominantly associated with persistent proteinopathy, other contributing factors can be leading a sedentary lifestyle, poor nutrition, and genetic predisposition. Irreversible amnesia is the most tragic case of Alzheimer's-related cognitive impairment. However, these cognitive impairments which do not have amnesia forms have their causes that can be reversed. Parkinson's disease comes second after Alzheimer's in terms of commonality as an impairment disorder. This arises because of dopamine depletion from striatum due to substantia nigra neuron loss. Oxidative stress, dysfunction mitochondria, cell–calcium imbalances, along with neuroinflammatory processes have been implicated in the pathophysiology of this disease. Polypharmacy, particularly in geriatric patients, can lead to substance-induced delirium which can have long-term impacts on cognition apart from direct brain injuries from accidents.

How Cognitive Impairment Is Evaluated

This is where we may want to consider an all-purpose mental capacity yardstick. We are talking about mental tests whose criteria are used by physicians to determine if there may be some cognitive decline at all ages or not (Figure 3.4). Many tools are now used in diagnosing dementia, including blood screening, Positron emission tomography (PET) scan, and cerebrospinal fluid analysis among other things. PET scans can show us how well our brain functions, while the cerebrospinal fluid has markers for Alzheimer's disease proteins. Intrathecal infusion has been proposed for administration of drugs, which are unable to cross the

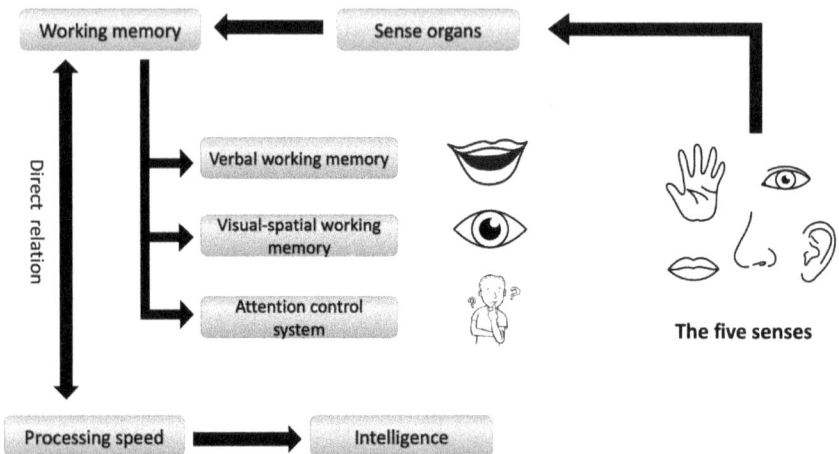

Figure 3.4   The cognitive process explained.

blood–brain barrier because it lacks its own blood supply (metabolic isolation). There is no cure for Alzheimer's disease and its symptoms worsen day by day as new treatments are being researched. These tests measure different cognitive functions, encompassing memory, orientation, concentration, language skills, and reasoning abilities among many others, though their effectiveness varies depending upon the individual patient's state of health.

Another integral part of evaluation is assessment for dementia risk factors which are modifiable or non-modifiable as knowledge of these will enable early intervention before the onset of debilitating manifestations leading to a great deal of independence lost over time. Mayo Clinic Brain Bank, National Institutes of Health (NIH), and Alzheimer's human brain bank are a few examples where researchers collaborate on tissue samples obtained during postmortem examinations to throw more light on the pathophysiology and potential breakthrough in medical treatment.

The MoCA is helpful to diagnose mild cognitive impairment in cases such as post-ischemic stroke or transient ischemic attack (TIA). These seven cognitive subtests make up the MoCA—visuo-executive, naming, attention, language, abstraction, delayed recall, and orientation. Basically, the Mini-Cog is some kind of brief neuropsychological assessment which only takes less than 3 minutes of your time when administering it. It involves a recall test and a clock drawing test (CDT). A score of −2 indicates the presence of dementia, while a score of 3–5 shows that there is no dementia present. Enter the modern era with digital tools like the Cambridge Neuropsychological Test Automated Battery (CANTAB) and the NIH Toolbox. CANTAB is employed to test neurocognitive dysfunctions associated with neurologic disorders, pharmacologic manipulations, and neurocognitive syndromes. NIH Toolbox cognitive battery measures different things like working memory and executive function among others. It can

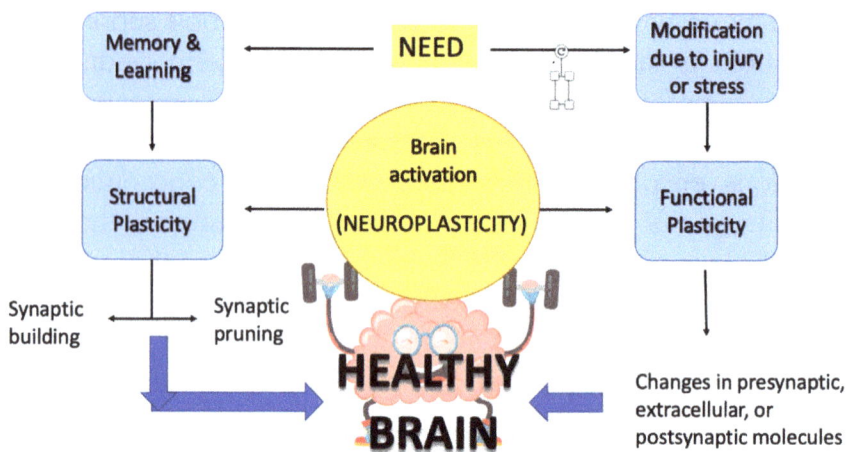

Figure 3.5   Brain health as a determinant of cognition.

be used on a tablet or iPad since it is run by software that gives results digitally (Figure 3.5).

With this understanding, we will be able to develop better tools for assessing brain health and maintaining a sound mind at all times. So, keep challenging your brain and stay sharp!

Hearing

In the auditory system, hearing is categorized as a sound perception process. This is when sound energy moves from one medium to another by vibrating the molecules present in these media (Figure 3.6). Have you ever thought of how the endless symphony of sounds in the world finally reaches your conscious awareness? Picture sound as a rockstar pulsing with energy vibrating air particles This rockstar travels through your external auditory meatus (ear canal) like an eager fan rushing to the front row. But there's still more to come! Our rockstar proceeds into our middle ear where it gets some groove going before finally hitting the stage—auditory cortex (located in temporal lobe).

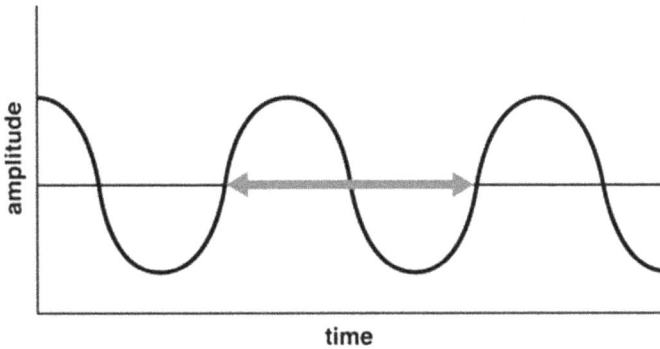

Figure 3.6   Sound waves traveling over time.
(National Institute of Health (USA), 2007).

Here is where magic takes place; the auditory cortex acts as a sound engineer decoding every wave of sound that comes into it based on three main factors:

Loudness: How intense is the rock star's performance? Is it a whisper or a roar?
Frequency: What note is the rockstar hitting on whether it is high- pitched and squeaky or low booming one?
Direction: Where is this rockstar performing on the stage? Left- or right-hand side?

These inputs come out as different sound experiences, which include cheering crowds, whispering conversations, murmurs, and sweet tune songs. In the same way that technical issues can spoil a concert, impaired hearing may be caused by damage along the auditory pathway (from ear canal to the brain). The process of accurate auditory signal transduction (for example, in spatial hearing or speech perception) relies on the precise integration of spectral (frequency-based) and temporal (time-based) components along this pathway (Figure 3.7). Such kind of damage at any level can result into hearing loss.

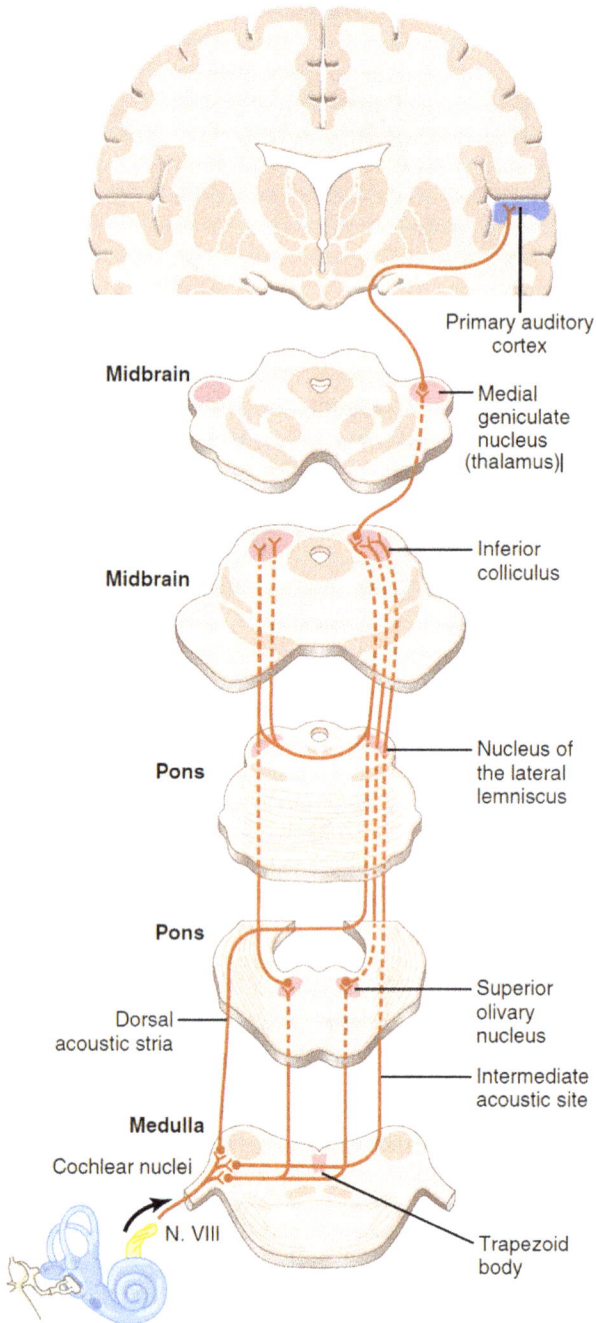

Figure 3.7   The central auditory pathway.

Link with Brain Health

An effective hearing system is like a cherished companion of the brain. Whenever our ears get damaged, the auditory regions of our brain might start performing the same function as the visual cortex—after all it has got resilience. That is because it is usual for reorganization, or recruitment of auditory cortical areas for visual processing due to functional plasticity. This is a normal healing phenomenon in case of hearing loss. But beware! In case hearing loss causes degeneration within the auditory pathway and the brain, then the brain cells may opt to change profession from sound detectors to visual ones. Keeping the sound signals crisp and clear is the main task folks. A fault anywhere along this sound pathway would lead to a malfunctioning hearing system, which is how closely brain health and hearing are related.

Impaired Hearing on Brain Health/Body Effect

As for tinnitus, it can be described as a stranger buzzing in the ear who has overstayed their welcome. It comes in two flavors: central and peripheral. Peripheral tinnitus is somewhat like an irritating low-pitched roar which refuses to go away often as a result of hyperexcitability of the cochlear nerve for low-frequency sounds. Central tinnitus, on the other hand, messes with the stimulus after it leaves the cochlea, all the way up to our brain's cortex. And loud noises are usually the culprits here.

Another stumbling block in the path of healthy hearing is the cochlear cell death triggered by loud noises causing loss of hearing. Other triggers include either mechanical faults, blood supply problems, or too much fun in the ears.

Diseases with Impaired Hearing

Fundamentally, hearing loss could be classified into sensorineural versus conductive types. Sensorineural deafness means that the inner ear or path leading to the auditory cortex has gone astray, whereas conductive deafness is about issues concerning the external ear and middle ear infections. Infections like varicella-zoster and Epstein–Barr virus have been known to carry out this offense as well as conditions like otitis and otosclerosis among others. Therefore, we need to always be observant and take care of these common offenders—like wax plugs that block sound waves traveling toward our ears. Thus, cleanliness of ear is an important factor for maintaining good hearing ability.

Measurement

Measuring hearing is like tuning your radio to get the clearest signal. These methods include pure tone audiometry (PTA) and brain-evoked response audiometry (BERA), which serve as standards when considering hearing loss. In other words, PTA refers to a subjective method of hearing evaluation based on patient's responses to given stimuli. It is noninvasive and provides accurate results but only when subjects respond accordingly. BERA is an objective method—it uses electrodes attached to the scalp through which signals are picked up by a computer (automated DJ) that is suitable for pediatric cases with complicated issues.

Visual Functionality

Have you ever wondered why you forget where you left your eyeglasses, even though they were only resting on your nose a moment ago? It is not just forgetfulness due to old age (though that can also happen!). The truth is, think of vision as Batman and Robin; best when together! Visual functions, such as visual acuity and cognition, together make up visual function. This advanced course is not just depending on occipital cortex; it incorporates different brain parts in a local and global scale.

Think of your brain as a huge movie theater. That's the camera sending pictures out of your eyes along those two highways called the optic nerves to the projector room located at the back of your head known as the occipital cortex. But here is an interesting fact: Even if your eyesight is said to be perfect (20/20), this doesn't mean that the whole film will be okay because in case there is dust or the projector doesn't work well, then no one will get to see clearly or at all. It is when one's overall health suffers. For example, visual cortex is affected following the decline of visual fields among other deteriorating brain faculties such as balance, hearing, or touch sensation, some getting enhanced while others getting deranged.

Central Mechanisms of Visual Function

Sabel and his colleagues introduced the "brain–eye–vascular triad," which is like the three musketeers of vision. It underscores the importance of cortical actions on vision. This might be accompanied by raised intracerebral pressure, undermined top-down modulation (attention, cognition), while emotionally triggered stress responses may affect blood vessels' regulation. An example is the autonomic nervous system which operates unconsciously over modalities in our bodies, among them heart activities such as pumping, kidney function, and digestion processes are a few. Supply of the blood to the eyes becomes inadequate when there is adverse effect on heart rate by compromised autonomic regulation; the vision becomes even weaker due to poor supply of fuel needed by eyes through blood vessels located within and behind them, respectively, in case pressure drops. Such kind of a response can also occur secondary to emotional disturbance be it excitement or melancholy. It is like a dance between your brain, eyes, and heart, where maintaining the rhythm is the key to a clear sight.

Link with Brain Health

Your brain interprets visual stimuli much like it does anything else—just practice regularly to keep those skills sharp! Sometimes, when you have been seeing unclearly for some time (from wearing an eye patch or due to a lazy eye), your mind starts forgetting how to see normally because after all every skill needs to be nurtured. It is a muscle that has to be put to use to make sense! Imagine if you didn't move your legs, how would you expect them to function well? Similarly, how can your brain decode any part of auditory information without processing it?

In childhood, our brains form excess connections among neurons. As we mature, these connections are "pruned" away, leaving the most used and efficient ones. When the eyes do not give a clear signal to the brain (blurry vision), these connections in the visual cortex are not solidified. Consequently, the brain's ability to process visual information from the weaker eye is permanently impaired. Therefore, as this happens, the brain makes more use of the other eye which has better vision. This can interfere with the working of connections from the weaker eye and thus it might become a perpetual recurrence—like a vicious cycle. Essentially, this is akin to having a beloved pair of shoes—you would wear them constantly disregarding others, leading to their falling apart.

Consequences of Visual Impairment

Impaired vision has behavioral and neural consequences. For example, people with impaired vision often have trouble judging distances because their depth perception is reduced. This impairs their ability to participate in normal physical activities like playing ball games or moving up/downstairs. Usually, when the brain and the eyes work harder for this compensation, headaches and eye fatigue ensue. Since vision plays a crucial role in coordinating movements, impaired vision can lead to clumsiness and difficulty performing tasks that require hand–eye coordination. Social events or activities where clear vision is needed may cause amblyopic individuals to obviate taking part in sports or games due to fear of underperforming.

When one eye is receiving blurry or weak signals for an extended period (as in amblyopia), the brain weakens the connections between that eye and the visual cortex. This can lead to permanent vision problems even if the eye itself is healthy. During critical periods of brain development in childhood, good quality vision is essential for proper wiring and organization of the visual system. Impaired vision can disrupt this process, leading to long-term problems with visual processing. The brain has an amazing ability to adapt and change (plasticity). However, prolonged visual deprivation can decrease the brain's plasticity, making it harder to recover from vision problems later in life. Research suggests that impaired vision might have a ripple effect, impacting other brain areas and functions that rely on visual information, such as spatial reasoning, attention, and memory.

Diseases damaging the occipital cortex can lead to "cortical visual impairment," also known as "cortical blindness" or "cerebral amblyopia." The culprits can be several.

Strokes or blockages in blood vessels supplying the occipital lobe can starve the brain of oxygen and nutrients, leading to damage. Infections like meningitis or encephalitis can inflame the brain, affecting the occipital cortex. Severe drops in blood sugar can temporarily impair brain function, including vision. Abnormal electrical activity in the brain during seizures can disrupt visual processing. Tumors growing in the occipital lobe can put pressure on the brain tissue, causing cortical blindness. Infections within the brain itself can damage the occipital cortex.

Diseases that affect vision, such as glaucoma (think of it like sly thief who steals peripheral vision) or cortical blindness which means your brain forgot how to see in totality, could be signs of bigger brain health issues. Glaucoma affects the peripheral fields; it also affects the optic nerve. It is like a red flag waving in the wind. Ignoring blurry vision could be like ignoring a flickering light on your car's dashboard; thereafter something bigger might be wrong with this. Color blindness is another disease associated with impaired visual function. It can be due to loss/reduction of red, blue, or green color pigments or any combination of these in the retina. An early diagnosis and treatment are essential for preventing further visual impairment.

Measurement

Testing visual function is like mapping the boundaries of your sight. Visual fields examination is about how much can one see without moving the eyeballs. Commonly, visual fields are examined with a fixed eye. This test may either be static or kinetic. The visual fields are more limited on the nasal and superior sides than on the lateral sides. This subjective measure of the field of vision is done in all four quadrants and gives an insight into its integrity.

The details and intricacies of visual acuity tests are such that you need specific examples of letters from across a distance, as shown by Snellen's chart. Visual acuity is the ability of the eyes to see high contrast of objects in terms of resolution. Usually its evaluation uses subjective methods with symbols and letters. Snellen's chart determines visual acuity by d/D formula where d is the distance at which the patient views the chart and D is the distance at which a person with normal vision would view it. The Ishihara chart is the ultimate color vision spy, uncovering hidden surprises in a sea of dots (Figure 3.8).

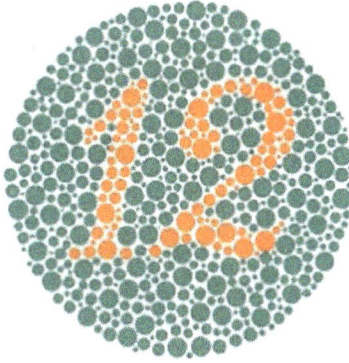

Figure 3.8   Ishihara chart.

Bibliography

1. Nall R. Brain Damage: What You Need to Know. Healthline, 2021. Available at https://www.healthline.com/health/brain-damage#outlook (accessed August 9, 2022).
2. Traumatic brain injury. Johns Hopkins Medicine, 2022. Available at https://www.hopkinsmedicine.org/health/conditions-and-diseases/traumatic-brain-injury(accessed August 9, 2022).
3. What are the possible effects of traumatic brain injury (TBI)? NIH, 2022, p. 104. Available at https://www.nichd.nih.gov/health/topics/tbi/conditioninfo/effects#(accessed August 9, 2022).
4. Lloyd D. Scanning the neurocracy: What do Brodmann areas do? Lloydia 1987;11:23–78. Available at http://commons.trincoll.edu/dlloyd/research/videos-web-pages/scanning-the-neurocracy-what-do-brodmann-areas-do/ (accessed December 20, 2021).
5. Ferng A. Brodmann areas. KENHUB, 2021. Available at https://www.kenhub.com/en/library/anatomy/brodmann-areas (accessed August 10, 2021).
6. Krausz AD, Korley FK, Burns MA. The current state of traumatic brain injury biomarker measurement methods. Biosensors. 2021;11(9):1–22.

7. Functional specialization. Encyclopedia of Evolutionary Psychological Science (pp. 3297–3297). Springer, 2021. Available at https://en.wikipedia.org/wiki/Functional_specialization_ (brain) (accessed December 21, 2021).

8. Lo RY. Longitudinal change of biomarkers in cognitive decline. Arch. Neurol. 2011;68(10):1257. doi: 10.1001/archneurol. 2011.123.

9. Boyle PA, Wang T, Yu L, Wilson RS, Dawe R, Arfanakis K. et al. The "cognitive clock": A novel indicator of brain health. Alzheimers Dement. 2021;17(12):1923–1937. doi: 10.1002/ alz.12351

10. Huang K, Ma X, Song R, Rong X, Li Y. Neurocomputing autonomous cognition development with lifelong learning: A self-organizing and reflecting cognitive network. Neurocomputing 2020;421:66–83. doi: 10.1016/j. neucom.2020.09.027.

11. Chai WJ, Hamid AIA, Abdullah JM. Working memory from the psychological and neurosciences perspectives: A review. Front. Psychol. 2018;9:1–16. doi: 10.3389/fpsyg.2018.00401.

12. Moriya J. Interactive effects of trait and state anxiety on visual spatial working memory capacity. Psychologia 2020;62:29–45. doi: 10.2117/psysoc.2020-B003.

13. Engle RW. Working memory and executive attention: A revisit. Perspect. Psychol. Sci. 2018;13(2):190–193. doi: 10.1177/1745691617720478.

14. Shipstead Z, Lindsey DRB, Marshall RL, Engle RW. The mechanisms of working memory capacity: Primary memory, secondary memory, and attention control. J. Mem. Lang. 2014;72(1):116–141. doi: 10.1016/j.jml.2014.01.004.

15. Schubert A, Frischkorn GT. Neurocognitive psychometrics of intelligence: How measurement advancements unveiled the role of mental speed in intelligence differences. Curr. Dir. Psychol. Sci. 2020;29(2):140–146. doi: 10.1177/0963721419896365.

16. Carlozzi NE, Beaumont JL, Tulsky DS, Gershon RC. The NIH toolbox pattern comparison processing speed test: Normative data. Arch. Clin. Neuropsychol. 2015;30:359–368. doi: 10.1093/ arclin/acv031.

17. Chiappe D, MacDonald K. The evolution of domain-general mechanisms in intelligence and learning. J. Gen. Psychol. 2005;132(1):5–40. doi: 10.3200/GENP.132.1.5-40.

18. Adolph K, Robinson SR. Motor development. In: Liben LS, Müller U, Lerner RM (Eds.). Handbook of Child Psychology and Developmental Science (7th ed., pp. 113–147). John Wiley & Sons, Inc., 2015. doi: 10.1002/9781118963418.childpsy204.

19. Neuroplasticity 101. BrainFutures, 2022. Available at https://www.brainfutures.org/neuroplasticity-101/ (accessed August 31, 2022).

20. Oberman L, Pascual-Leone A. Changes in plasticity across the lifespan. Prog. Brain Res. 2013;207:91–120. doi: 10.1016/B978-0-444-63327-9.00016-3

21. Cognitive disorders. Mount Sinai. 2022. Available at https://www.mountsinai.org/care/neurology/services/cognitive-disorders (accessed August 31, 2022).

22. Cognitive impairment: A call for action, now! Centers for Disease Control and Prevention, 2011. Available at https://www.cdc.gov/aging/pdf/cognitive_impairment/cogimp_poilicy_final.pdf

23. Jongsiriyanyong S, Limpawattana P. Mild cognitive impairment in clinical practice: A review article. Am. J. Alzheimer's Dis. Other Demen. 2018;33(8):500–507. doi: 10.1177/1533317518791401.

24. Aarsland D, Batzu L, Halliday GM, Geurtsen GJ, Ballard C, Ray Chaudhuri K. et al. Parkinson disease-associated cognitive impairment. Nat. Rev. Dis. Prim. 2021;7(1):47. doi: 10.1038/s41572-021-00280-3

25. Dhakal A, Bobrin BD. Cognitive deficits. Treasure Island, FL: StatPearls, 2022. Available at https://www.ncbi.nlm.nih.gov/books/NBK559052/ (accessed August 31, 2022).

26. Mini-mental state examination. Physiopedia, 2021. Available at https://www.physio-pedia.com/Mini-Mental_State_Examination (accessed December 3, 2021).

27. Cognitive testing. MedlinePlus. 2020. Available at https://medlineplus.gov/lab-tests/cognitive-testing/ (accessed August 31, 2022)

28. Arevalo-Rodriguez I, Smailagic N, Roqué-Figuls M, Ciapponi A, Sanchez-Perez E, Giannakou A., et al. Mini-Mental State Examination (MMSE) for the early detection of dementia in people with mild cognitive impairment (MCI). Cochrane Database Syst. Rev. 2021;7(7):CD010783. doi: 10.1002/14651858.CD010783.pub3

29. Cherry K. The Wechsler Adult Intelligence Scale. Verywellmind, 2020.
30. Cambridge Neuropsychological Test Automated Battery (CANTAB). 2001. Available at https://www.cdc.gov/me-cfs/pdfs/wichita-data-access/cantab-doc.pdf
31. Hessl D, Sansone SM, Berry-Kravis E, Riley K, Widaman KF, Abbeduto L., et al. The NIH toolbox cognitive battery for intellectual disabilities: Three preliminary studies and future directions. J. Neurodev. Disord. 2016;8(1):35. doi: 10.1186/s11689-016-9167-4.
32. National Institute of Health (US). Information about Hearing, Communication, and Understanding. NIH Curriculum Supplement Series. Bethesda (MD): Biological Sciences Curriculum Study, 2007. Available at http://www.ncbi.nlm.nih.gov/books/NBK20366/#A547
33. Pilati N, Ison MJ, Barker M, Mulheran M, Large CH, Forsythe ID, et al. Mechanisms contributing to central excitability changes during hearing loss. Proc. Natl. Acad. Sci. 2012;109(21):8292–8297. doi: 10.1073/pnas.1116981109.
34. Johnson JCS, Marshall CR, Weil RS, Bamiou D-E, Hardy CJD, Warren JD. Hearing and dementia: From ears to brain. Brain 2021;144(2):391–401. doi: 10.1093/brain/awaa429.
35. Campbell J, Sharma A. Cross-modal re-organization in adults with early stage hearing loss. PLoS One 2014;9(2):e90594. doi: 10.1371/journal.pone.0090594.
36. Burry M. Understanding auditory deprivation: Why untreated hearing loss is bad for your brain. 2021. Available at https://www.healthyhearing.com/report/46306-Hearing-loss-auditory-deprivation (accessed December 3, 2021).
37. Yin TCT, Smith PH, Joris PX. Neural mechanisms of binaural processing in the auditory brainstem. Compr. Physiol. 2019; 9:1503–1575. doi: 10.1002/cphy.c180036
38. Noreña AJ. Revisiting the cochlear and central mechanisms of tinnitus and therapeutic approaches. Audiol. Neurotol. 2015;20(Suppl. 1):53–59. doi: 10.1159/000380749.
39. Basta D, Gröschel M, Ernst A. Central and peripheral aspects of noise-induced hearing loss. HNO, 2018;66(5):342–349. doi: 10.1007/s00106-017-0442-9.
40. Mcgrath A. Types of hearing loss. John Hopkins Medicine. 2022. Available at https://www.hopkinsmedicine.org/health/

conditions-and-diseases/hearing-loss/types-of-hearing-loss (accessed September 18, 2022).

41. Schreiber BE, Agrup C, Haskard DO, Luxon LM. Sudden sensorineural hearing loss. Lancet 2010;375(9721):1203–1211. doi: 10.1016/S0140-6736(09)62071-7.

42. Hill-Feltham PR, Johansson ML, Hodgetts WE, Ostevik AV, McKinnon BJ, Monksfield P, et al. Hearing outcome measures for conductive and mixed hearing loss treatment in adults: A scoping review. Int. J. Audiol. 2021;60(4):239–245. doi: 10.1080/14992027.2020.1820087.

43. Walker JJ, Cleveland LM, Davis JL, Seales JS. Audiometry screening and interpretation. Am. Fam. Physician 2013;87(1):41–47.

44. Kim D, Kwon J, Kim J-Y, Cha H-S, Kim Y-W, Kim I, et al. New method for pure-tone audiometry using electrooculogram: A proof-of-concept study. Sensors. 2018;18(11):3651. doi: 10.3390/s18113651

45. Abhilash AM, Saritha HM. BERA as a screening tool for evaluating cases of SNHL due to suspected retrocochlear pathology. Int. J. Adv. Res. Med. 2021;3(2):10–14. doi: 10.22271/27069567.2021.v3.i2a.208.

46. Rovner BW, Casten RJ, Massof RW, Leiby BE, Tasman WS. Psychological and cognitive determinants of vision function in age-related macular degeneration. Arch. Ophthalmol. 2011;129(7):885–890. doi: 10.1001/archophthalmol.2011.146.

47. Sabel BA, Flammer J, Merabet LB. Residual vision activation and the brain–eye–vascular triad: Dysregulation, plasticity and restoration in low vision and blindness—A review. Restor. Neurol. Neurosci. 2018;36(6):767–791. doi: 10.3233/RNN-180880.

48. Kasi A, Faiq M, Chan K. In vivo imaging of structural, metabolic and functional brain changes in glaucoma. Neural. Regen. Res. 2019;14(3):446. doi: 10.4103/1673-5374.243712.

49. Purves D, Augustine G, Fitzpatrick D (Eds.). Central visual pathways. In: Neuroscience (2nd ed.). Sunderland, MA: Sinauer Associates, 2001. Available at https://www.ncbi.nlm.nih.gov/books/NBK11034/.

50. Levi DM. Rethinking amblyopia 2020. Vision Res. 2020;176:118–129. doi: 10.1016/j.visres.2020.07.014.

51. Hoyt CS. Visual function in the brain-damaged child. Eye. 2003;17(3):369–384. doi: 10.1038/sj.eye.6700364.

52. Papageorgiou E, Asproudis I, Maconachie G, Tsironi EE, Gottlob I. The treatment of amblyopia: Current practice and emerging trends. Graefe's Arch. Clin. Exp. Ophthalmol. 2019;257(6):1061–1078. doi: 10.1007/s00417-019-04254-w.

53. Jonauskaite D, Camenzind L, Parraga CA, Diouf CN, Mercapide Ducommun M, Müller L, et al. Colour-emotion associations in individuals with red-green colour blindness. PeerJ. 2021;9:e11180. doi: 10.7717/peerj.11180

54. Schuster AK, Erb C, Hoffmann EM, Dietlein T, Pfeiffer N. The diagnosis and treatment of glaucoma. Dtsch. Arztebl. Int. 2020;117(13):225–234. doi: 10.3238/arztebl.2020.0225.

55. Reddy GR. Visual fields in glaucoma: An overview. J. Ophthalmol. Clin. Res. 2021;1(1):38–54. doi: 10.4103/jocr.jocr_22_21.

56. Hamilton R, Bach M, Heinrich SP, Hoffmann MB, Odom JV, McCulloch DL, et al. VEP estimation of visual acuity: A systematic review. Doc. Ophthalmol. 2021;142:25–74. doi: 10.1007/s10633-020-09770-3.

57. Ishihara test. Webeyeclinic. 2022, pp. 5–6. Available at https://www.medicalonline.pk/index.php?route=product/product&product_id=897(accessed September 19, 2022).

Chapter 4

Determinants of Brain Health

Introduction

In continuation of the previous chapter, several other factors like motor function, emotional well-being, and age can also affect brain's functioning. Different functions/faculties of the brain corresponding to various brain areas can be quantified. Several disorders like motor neuron disease and depression can pose a serious threat to its integrity. Biomarkers also have a share in the cake depicting brain health. A biomarker can be used for diagnostic or prognostic purposes, to check the response to treatment, and even to find out the etiological process behind the condition.

Motor Function

Evading wild pigeons while out for a morning stroll on foot, to remembering the grocery list—the mind is the operational force behind them all. But how on earth does this 3-pound

DOI: 10.1201/9781032676579-4

blob of tissue orchestrate all these amazing feats? Let's delve into our brain's response nexus—the motor system!

Imagine your brain as the control center for a giant robot (that robot being your amazing body, of course), think about it receiving constant sensory inputs—such as what we get from spy cameras in real time! Did you just notice that delicious doughnut across the street? Boom! The brain sends a signal to the salivary glands. Feeling that wobbly table beneath you? The brain commands the posture muscles to adjust. This way, information and action flow in a constant stream fashion (Figure 4.1).

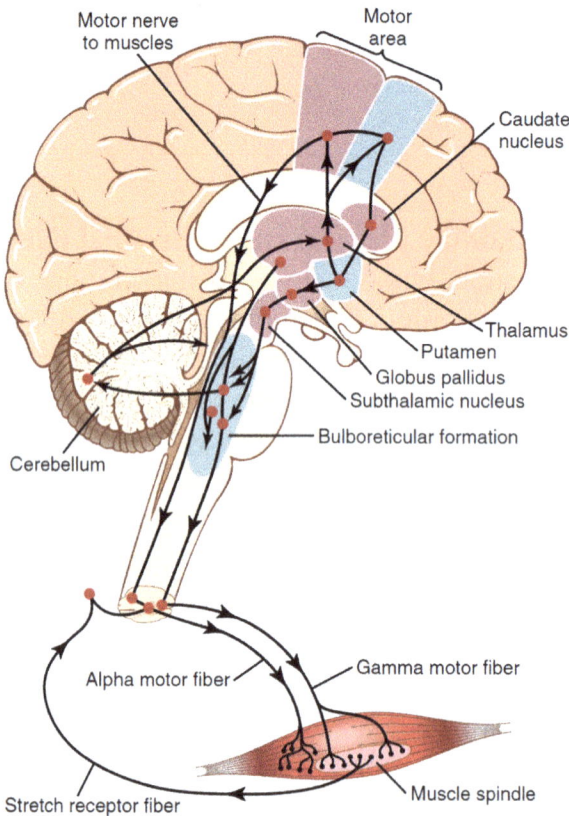

Figure 4.1   Pathways of motor function between the skeletal muscles and the nervous system.

But here's the twist: Your brain does not command muscles directly like a marionette. Rather, it employs a series of relay stations that have specific functions:

1. *Spinal Cord:* It resembles the central nervous system of robot, exchanging messages between it and the body's "muscles and motors," in other words, your muscles.
2. *Reticular Substance:* This can also be seen as an innate driving force of your mind, the built-in "get moving" center. It initiates automatic movements like walking or keeping the balance—just like the autopilot function of a robot.
3. *Basal Ganglia:* In other words, they are like fine-tuning knobs on a robot. These structures help in smooth well-controlled movements that prevent many incidences of tripping over your feet while walking (most times!).
4. *Cerebellum:* Think about it as the balance coach for a robot. It helps coordinate accurate moves and maintain your upright posture all the time!
5. *Motor Cortex:* This is the grand master, the boss of motion. It sits majestically atop your head sending last-minute instructions to every muscle in your body directing what needs to be done at which particular moment.

All in all, your brain resembles a melodious orchestra where each movement is meticulously planned and executed. Keeping you as mobile as a highly oiled up motor car!

Have you ever wondered about how you tie your shoelaces, brush your teeth, or even execute that complicated dance move you once watched online? It comes down to a three-way team effort: yourself, your body, and the environment itself!

You might be wondering now if your body is the MVP (most valuable player) in this regard. I mean, it is the muscles that do most of the hard work, right? However, research asserts that it is your brain. Think of it as the quarterback calling the plays—without clear instructions, your body (the receivers and running backs) wouldn't know what to do!

Now comes the cool part: Your mind is not on its own. It often employs a technique called motor imagery, which is basically akin to mental rehearsal. You can imagine that you are doing a push-up before even beginning exercise. This helps to bridge the gap between thinking about a movement and actually making it happen. It is like giving your body a heads-up regarding what to expect next.

For a well-performed action, there has to be more than just an intelligent preparation process that goes through the mind. Balance, strength, dexterity (fancy word for smooth moves), endurance, and visual–motor coordination also come in handy. Luckily for us, the motor cortex of the brain plays this coordinating role evident through the smoothness and efficiency of movements (Figure 4.2).

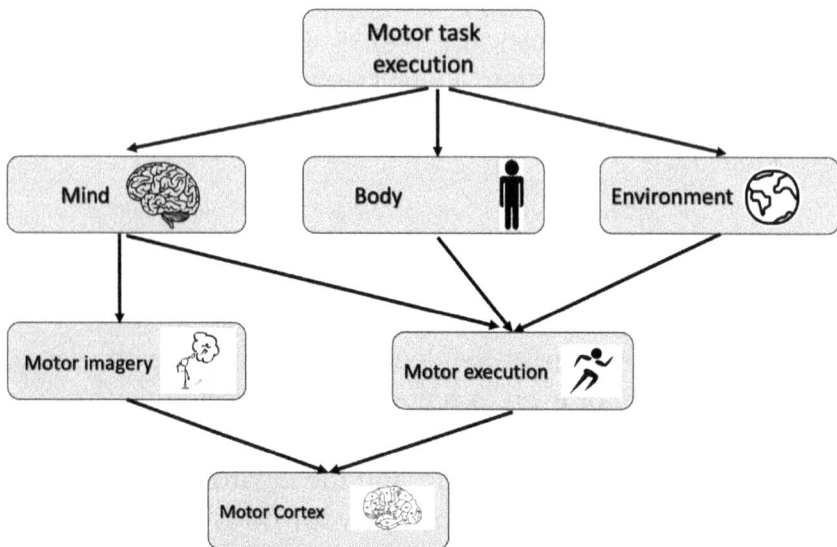

Figure 4.2 Process of motor execution.

Different aspects of movement are governed by different regions of your brain or other parts even outside it! It therefore takes the combined effort of strength and balance between motor cortex, cerebellum, and basal ganglia to achieve motility with sensitivity. Just like different departments in a company, each brain region has its own specialty!

Thus, when you next perform something as simple as taking a breath, recognize what an incredible collaboration there is between your mind, body, and even environment at large. It is a testament to the amazing adaptability and teamwork of your brain—the ultimate control center for your daily life!

Central Mechanisms of Motor Functions

The brain is divided into various divisions that are related to controlling the body's activities. Through two pathways, the central nervous system interacts with the world outside us: sensory pathways allow it to gather information, while output pathways enable it to send commands to muscles and glands.

Planning of motor activities occurs in the secondary motor areas like supplementary motor area, premotor cortex, as well as posterior parietal cortex. Primary motor cortex determines what you're going to do (signs off on actions), whereas it lets others plan everything else such as secondary areas or cerebellum. Supplementary motor cortex prepares complex sequence of movements; premotor cortex ensures that our body and arms are oriented correctly toward targets.

Link with Brain Health

Even a little exercise can keep the brain fit as a fiddle. Messenger proteins called myokines play messengers between the brain, muscles, and body tissues. Undeniably, these myokines help foster brain health by facilitating formation of new neurons, thereby increasing learning ability and ability to remember things well. With regular exercise, one can lower chances of getting brain-related diseases such as Alzheimer's and Parkinson's. On the other hand, myokines regulate overall body metabolism in addition to muscle metabolism ensuring that all organs move and work at the same pace. Interleukin 6 (IL-6) and brain-derived neurotrophic factor (BDNF) are myokines that help muscles adapt to exercise and boost brain performance, memory, and decision-making.

Consequences of Impaired Motor Function

When the primary motor cortex is impaired, secondary motor areas such as premotor cortex and supplementary motor cortex take over. They directly activate spinal cord motor neurons ensuring completion of all tasks. Studies on rats show that when neurons surrounding the injured site become hyperactive in response to trauma, they release growth factors that facilitate recovery. This involves generating new synapses and utilizing the undamaged neurons for recovery. At the later stage of injury period, the brain produces a protein called GAP-43. It acts like a brain fertilizer promoting sprouting of new connections among the injured neurons. It is basically your brain flexing its muscles and saying, "We're gonna grow our way out of this!" Scientists saw a spike in another molecule called synaptophysin. This is like the brain's way of building new communication bridges between neurons. It is all about creating new pathways to get those movement messages flowing again. Picture the head injury area as a very excited fan club. And such cells become abnormally excitable, overworking themselves to transmit information. However, our brain knows how to put everything under control every time. It modifies some receptors to tone them down and prevents saturation of the entire system.

Post Injury: A Hub of Regeneration

The helper neurons release growth factors into this area surrounding the injury—just like a group of medics who start reviving a fallen person.

The truth is, after an injury, some motor functions might be permanently lost. But don't worry; there are some functions which can be reassigned to those areas serving as helpers while others are undergoing recovery. This is evidence of how powerful and relentless our brains are when it comes to getting you back to your best, one step (or thought!) at a time.

Diseases of Movement Dysfunction

When there is a dysfunction of motor cells, motor neuron diseases develop. Such ailments include amyotrophic lateral sclerosis (ALS), spinobulbar muscular atrophy, hereditary spastic paraparesis, and acute flaccid paralysis associated with viral infection.

Amyotrophic lateral sclerosis involves various types of upper and lower motor neuron lesions. It usually starts in the limbs; however, while some remain intellectually okay, as famously exemplified by Stephen Hawking who carried on with his physics thought-provoking research despite the diagnosis of ALS (Figure 4.3). Activated T lymphocytes and microglia also play significant roles in causing ALS through an inflammatory process taking place within the nervous system.

Parkinson's disease is another condition characterized by deranged motor functions triggered by α-synuclein protein which would make one experience tremors while walking slowly, and also movements become rigid. Another hereditary neurodegenerative disease that affects the motor system is Huntington's disease resulting from mutation on chromosome 4 where it presents with chorea, bradykinesia, balance, and

Figure 4.3   A photo of Stephen Hawking.

Photo by Makrem Larnout. Available at https://www.facebook.com/ photo.php?fbid=10160152171290613&set=t.100044191577876&type=3.

gait disturbance. Abnormal neurofilament light (NfL) protein levels serve as markers of a good prognosis for the disease.

Measurement

Proper execution of motor actions demands balance, strength, dexterity, endurance, and visual–motor coordination. A motor action can have all or a combination of these. Motor strength can be checked both subjectively and objectively and can be used as a good representative of motor function. Manual motor testing (MMT) easily gauges the muscle strength in a subjective way using one's hands, although it greatly depends on examiner's experience and expertise. Another method for assessing muscle strength is through the use of a handheld dynamometer. A dynamometer is a device that records the force applied by compressed muscles measured in kg/m^2 when using it as a base.

Emotional Well-Being

Imagine your brain as an electronic jukebox dishing out joyous odes, melancholic ballads, and angry punk anthems for your emotions. Everything boils down to your attitude, the quality of sleep you have, social support or not having, work stress, or those little things that stress you out like being asked to work on weekends.

What is emotional intelligence anyway?

According to Schutte et al., emotional intelligence is "the ability to adaptively perceive, understand, regulate, and harness emotions in the self and others." Simply put, it's the brain's way of saying, "I've got this!" In other words, emotional intelligence can be seen as the highway to emotional wellness. The two stars of this show are mood and self-esteem.

Positive Mood: Imagine yourself feeling active, energetic, and lively.

Negative Mood: Imagine feeling sad, tired, and lost.

Emotions have a great impact on a person's entire well-being, leading to further physical or mental disturbances. If a person is emotionally labile, he may fall victim to addictions and other dangerous pleasure-seeking pursuits which will eventually deteriorate his health further.

Feeling good about oneself is a brain's way of giving a pat on the back. When things are going well, there's this sense that you are on top of the world but when they aren't, it feels like a very scary place.

Central Mechanisms of Emotional Well-Being

Emotions serve as the brain's personal GPS by giving a sense of a person's internal state enabling him to lead a certain lifestyle. They act as adaptive tools in your brain that ensure you stick to your set objectives. Emotional regulation spans everything from neurons firing to how you interact socially. These include the amygdala, hypothalamus, thalamus, prefrontal cortex, and the sensory cortex interacting with various feedback processes in the body, principally hormones and proprioception.

When it comes to emotions in the brain, three major neuromodulators stand out: opioids, serotonin, and dopamine. These guys are the best of the best at boosting mood levels!

Though life may get hard, the autonomic nervous system as well as the HPA axis come to our rescue. They're the brain's all-star team for maintaining balance and handling stress. Equally important is immunity that uses cytokines such as tumor necrosis factor a (TNF-a) and IL-6 to maintain homeostasis within the body. All in all, it is an emotional–immunity loop; feelings influence immunity and vice versa in a very dynamic process.

Research has shown that major depressive disorder (MDD) is associated with reduced adaptive immunity where pro-inflammatory cytokines are coming into the party uninvited. This is when the immune system is certainly under duress!

The Brain–Body Connection: Emotion in Action

Plutchik, "the father of Emotion Theory," in 1984 provided a definition that has lasted up to now: "Emotion refers to reactions to an appropriately evocative stimulus involving cognitive appraisal/perception, subjectively experienced feeling, autonomic and neural arousal, expressive behavior, and goal-directed activity." Studies have shown that for any individual whose positive emotions have been induced through stimulating the brain, they are most likely to be satisfied by their jobs leading to increased workplace productivity and at the same time improve social interactions with others. In short, to keep your mental and physical gears turning, you need a healthy dose of good vibes.

When Emotions Wilt, So Does the Body

Feeling down can take a toll on your physical health. Stress and emotional wounds like depression and anxiety behave like a terrible weed in the garden of your heart that kills motivation and joy. This can create a vicious cycle where you lose interest in things that once gave pleasure, work becomes a struggle, and relationships suffer too. Remember Ebenezer Scrooge in *A Christmas Carol*? He was emotionally unwell and his entire body was affected (Figure 4.4).

Figure 4.4   A photo of statue of Ebenezer Scrooge.

Photo by Martijn Nijenhuis. Available at https://www.flickr.com/ photos/martijnnijenhuis/19177600396/ and https://www.flickr.com under CC BY 2.0 (https://creativecommons.org/licenses/by/2.0/).

The good news, however, is that just as Scrooge was eventually redeemed, taking care of your mental health will positively impact all aspects of life.

Diseases with Emotional Woes

The World Health Organization (WHO) has a host of mental disorders:

Depression: A neurobiological disorder related to the levels of serotonin ranging from bipolar to major depressive disorder. It has symptoms ranging from low mood to suicide ideation.

PTSD: It is a nasty beast that messes up with your self-organization, mood, self-concept, and relationship. Traumatic brain injury (TBI) affects 50 million people globally. Such patients have neuroinflammation, oxidative stress, and neuron chaos causing these individuals to have wars going on in their heads.

Schizophrenia: This disorder is characterized by hallucinations, delusions, and cognitive issues. The main culprit in schizophrenia is an excess of dopamine.

Autism Spectrum Disorder (ASD): It includes both genetic predisposition and environmental factors. Children with autism have difficulty relating to others, communicating effectively with speech or nonverbal expressions, or showing empathy toward others' plight; hence their behavior is perceived as being strange or even challenging in some cases. Autism etiology depends on several factors; but mostly it comes down to gene–environment interaction models during prenatal period. Developmental stage disorder associated with autism encompasses impairment social skills development leading to problems such as poor language use abilities among many others.

Anxiety: Panic and mental agitation affect roughly 6% of the population, suggesting that something triggers a "fight or flight" response in humans. This response often manifests through symptoms such as agitation, irritability, fatigue, and a sense of disconnection.

How to Measure Emotional Well-Being

You might want to consider taking the Mayer–Salovey–Caruso Emotional Intelligence Tests (MSCEIT), Self-report Emotional Intelligence Test (SREIT), and Trait Emotional Intelligence Questionnaire (TEIQue). This would depend on whether you are only feeling blues due to some recent happenings or there is a potential for long-term emotional traumas. And these questionnaires are also a game of expertise and repeated measures.

When your mind feels like testing its emotional smartness or you just want to keep your brain healthy, there are numerous tools for you, including the Short Form of Trait Emotional Intelligence Questionnaire (TEIQue-SF). It is amazing how this measure can be applied in so many different ways other than English!

Age

Types of Age and Its Measurement

Aging is classified into biological, psychological, and functional aging; it is a multifaceted process. Comfort wrote: "But what is being measured when we measure age is decrease in viability and an increase in vulnerability." Physical deterioration such as DNA damage (accumulation of somatic mutations), shortened telomeres length, elevated levels of reactive oxygen species (ROS), mitochondrial dysfunction, activation oncogenes, damaging epigenetic mechanisms, as well as other factors affect the aging process. Similarly, diseases (heredity and acquired) and predisposition to them serve as indicator points for its assessment.

Basically, chronological age tells us how long one has lived. However, when it comes to particular brain aspects comprising all mental processes, it may not adequately predict individual's health status.

Skeletal age is a standard means for measuring age and growth, in particular in children via using radiological means. These means include X-rays and magnetic resonance imaging (MRI) technique; recently, ultrasound has also been explored for this purpose as an alternative to high-radiation exposure risk.

Functional aging is all about how sturdy or fragile your body is when life throws curveballs. It is like checking the dashboard of a car for warning lights, with various indicators giving us the lowdown on what's happening under the hood.

The psychological consequences of aging are not limited to the cognitive realm, but extend into emotional processes like emotion recognition and response latency. In senescence, slow down of executive function, reduced capacity of working memory to hold and assimilate thoughts, events, and emotions are noted along with slowed processing speed. In decision-making at old age most of the times one needs external support for just making decisions because one's capacity goes down. In summary, it is fair to describe aging as a slow dance which involves more deliberation, while maneuvering between our emotions and reasoning. Boyle and his colleagues (2021) found out that age-related cognitive changes otherwise known as biological age are more predictive of health status than chronological age leading him to coin the phrase "Cognitive Clock." The Cognitive Clock takes into account cognition while indicating how healthy or unhealthy one's brain could be.

Effect of Aging on Brain Health

Brain health changes with advanced years much like an overdue renovation in an old building where the scaffolding (neurons and neurotransmitters) starts falling apart while wiring (synaptic connections) begins showing some wear. Even grip strength can provide a clue regarding how much someone's brain has shrunk over time while there is an increasing risk of neurological and psychiatric disorders following advancing age. In old age, the likelihood of psychiatric or neurodegenerative diseases is high. Diseases such as seizures, psychiatric ailments, stroke down to neurodegeneration are related to aging. Hence, aging is a nonmodifiable risk factor for disorders of the higher nervous centers, and as a person ages, onset of age-related neurodegenerative diseases such as Alzheimer's disease has a higher chance of manifestation. Also in old age, long-term brain disorders such as neurodegenerative diseases are associated with disruption of blood–brain barrier (BBB) and blood–cerebrospinal fluid barrier (BCSFB). Due to lack of sufficient safeguards in the brain, more leukocytes move across the barriers, leading to inflammation and release of inflammatory indicators. In addition, brain markers are being secreted into the bloodstream while their secretion and clearance are also affected. Therefore, aging should be viewed as an unyielding force against which our brains will be reshaped as long as life continues with researchers striving to extend this carnival indefinitely!

Biomarkers

Protein Biomarkers and Their Characteristics

Proteins that could serve as biomarkers resemble spies within body fluids like CSF, serum, urine, etc., waiting for an opportunity to leak secretively. Biomarkers also serve as informants to the healthcare practitioners when diagnosing conditions, predicting the outcomes, and even judging the efficacy of treatment protocols. Therefore, a good biomarker should be regarded as an inspector who is not only particular but also time-conscious, cheap, and highly efficient. They are predominantly found in the blood, making them easy to trap using a simple needle stick; yet some hide somewhere in the spinal cord making it difficult for doctors to access them. Simple needle pokes in the arm would catch many of them wandering around our bloodstreams, but others are more selectively restricted to CSF where detection becomes tricky. Thus, these detectives are very essential whenever somebody goes hunting for answers about their health status even if they sometimes make us work for it!

Bibliography

1. Reuben DB, Magasi S, McCreath HE, Bohannon RW, Wang YC, Bubela DJ, et al. Motor assessment using the NIH Toolbox. Neurology. 2013;80(Suppl 3):65–75. doi: 10.1212/WNL.0b013e3182872e01
2. Canedo A. Primary motor cortex influences on the descending and ascending systems. Prog Neurobiol. 1997;51(3):287–335. Doi: 10.1016/s0301-0082(96)00058-5
3. Sherwood L. Human Physiology from Cells to Systems. Ninth. Belmont CA; 2015; 319–321pp.
4. Scott SH. Inconvenient truths about neural processing in primary motor cortex. J Physiol. 2008;586(5):1217–1224. DOI: 10.1113/jphysiol.2007.146068.
5. How do strong muscles keep your brain healthy? MIT Technology Review, 2022. Available at https://www.technologyreview.com/2022/08/22/1057660/strong-muscles-healthy-brain/ (accessed September 15, 2022).
6. Kim S, Choi J-Y, Moon S, Park D-H, Kwak H-B, Kang J-H. Roles of myokines in exercise-induced improvement of neuropsychiatric function. Pflügers Arch.: Eur. J. Physiol. 2019;471(3):491–505. doi: 10.1007/s00424-019-02253-8.

7. Sharma N, Cohen LG. Recovery of motor function after stroke. Dev. Psychobiol. 2012;54(3):254–262. doi: 10.1002/dev.20508.

8. Nudo RJ. Mechanisms for recovery of motor function following cortical damage. Curr. Opin. Neurobiol. 2006;16(6):638–644. doi: 10.1016/j.conb.2006.10.004.

9. Foster LA, Salajegheh MK. Motor neuron disease: Pathophysiology, diagnosis, and management. Am. J. Med. 2019;132(1):32–37. doi: 10.1016/j.amjmed.2018.07.012.

10. Van Damme P, Robberecht W. Recent advances in motor neuron disease. Curr. Opin. Neurol. 2009;22(5):486–492. doi: 10.1097/WCO.0b013e32832ffbe3.

11. Endo F, Komine O, Yamanaka K. Neuroinflammation in motor neuron disease. Clin. Exp. Neuroimmunol. 2016;7(2):126–138. doi: 10.1111/cen3.12309.

12. Vázquez-Vélez GE, Zoghbi HY. Parkinson's disease genetics and pathophysiology. Annu. Rev. Neurosci. 2021;44(1):87–108. doi: 10.1146/annurev-neuro-100720-034518.

13. McColgan P, Tabrizi SJ. Huntington's disease: A clinical review. Eur. J. Neurol. 2018;25(1):24–34. doi: 10.1111/ene.13413.

14. Muscle strength testing. Physiopedia, 2021, p. 2. Available at https://www.physio-pedia.com/Muscle_Strength_Testing (accessed September 15, 2022).

15. Emotional wellness toolkit. US Department of Health and Human Services, 2018, p. 20892. Available at https://www.nih.gov/health-information/emotional-wellness-toolkit (accessed December 3, 2021).

16. Schutte NS, Malouff JM, Simunek M, McKenley J, Hollander S. Characteristic emotional intelligence and emotional well-being. Cogn. Emot. 2002;16(6):769–785. doi: 10.1080/02699930143000482.

17. Brennan D. What to know about emotional health. WebMD, 2021. Available at https://www.webmd.com/balance/what-to-know-about-emotional-health#:~:text=Emotional health is one aspect to a professional for help (accessed August 10, 2022).

18. Nyklíček I, Vingerhoets A, Zeelenberg M. Emotion regulation and well-being. In: Nyklíček I, Vingerhoets A, Zeelenberg M (Eds.). Paper Knowledge: Toward a Media History of Documents (Vol. 3, pp. 49–58). New York, NY: Springer, 2011. doi: 1007/978-1-4419-6953-8.

19. Arbib MA, Fellous JM. Emotions: From brain to robot. Trends Cogn. Sci. 2004;8(12):554–561. doi: 10.1016/j.tics.2004.10.004.
20. Pfau ML, Russo SJ. Peripheral and central mechanisms of stress resilience. Neurobiol. Stress 2015;1(1):66–79. doi: 10.1016/j.ynstr.2014.09.004.
21. Salim S. Oxidative stress: A potential link between emotional wellbeing and immune response. Curr. Opin. Pharmacol. 2016;29:70–76. doi: 10.1016/j.coph.2016.06.006.
22. Borod JC. Interhemispheric and intrahemispheric control of emotion. J. Consult. Clin. Psychol. 1991;60(3):339–348. doi: 10.1037//0022-006x.60.3.339.
23. Burgdorf J, Panksepp J. The neurobiology of positive emotions. Neurosci. Biobehav. Rev. 2006;30(2):173–187. doi: 10.1016/j.neubiorev.2005.06.001.
24. Morris RG, Morris LW, Britton PG. Factors affecting the emotional wellbeing of the caregivers of dementia sufferers. Br. J. Psychiatry. 1988;153:147–156. doi: 10.1192/bjp.153.2.147.
25. Karol E, Smith D. Impact of design on emotional, psychological, or social well-being for people with cognitive impairment. Heal. Environ. Res. Des. J. 2019;12(3):220–232. doi: 10.1177/1937586718813194.
26. World Health Organization. Mental disorders. 2022. Available at https://www.who.int/news-room/fact-sheets/detail/mental-disorders
27. Benazzi F. Various forms of depression. Dialogues Clin. Neurosci. 2006;8(2):151–161. doi: 10.31887/DCNS.2006.8.2/fbenazzi.
28. Moncrieff J, Cooper RE, Stockmann T, Amendola S, Hengartner MP, Horowitz MA. The serotonin theory of depression: A systematic umbrella review of the evidence. Mol Psychiatry. 2023;28:3243–3256. doi: 10.1038/s41380-022-01661-0
29. Heim E, Karatzias T, Maercker A. Cultural concepts of distress and complex PTSD: Future directions for research and treatment. Clin. Psychol. Rev. 2022;93:102143. doi: 10.1016/j.cpr.2022.102143.
30. Monsour M, Ebedes D, Borlongan CV. A review of the pathology and treatment of TBI and PTSD. Exp. Neurol. 2022;351:114009. doi: 10.1016/j.expneurol.2022.114009.
31. Rubinov M, Bullmore E. Schizophrenia and abnormal brain network hubs. Dialogues Clin. Neurosci. 2013;15(3):339–349. doi: 10.31887/DCNS.2013.15.3/mrubinov.

32. Reichenberg A. The assessment of neuropsychological functioning in schizophrenia. Dialogues Clin. Neurosci. 2010;12(3):383–392. doi: 0.31887/DCNS.2010.12.3/areichenberg.
33. Zeidan J, Fombonne E, Scorah J, Ibrahim A, Durkin MS, Saxena S, et al. Global prevalence of autism: A systematic review update. Autism Res. 2022;15(5):778–790. doi: 10.1002/aur.2696.
34. Chaste P, Leboyer M. Autism risk factors: Genes, environment, and gene–environment interactions. Dialogues Clin. Neurosci. 2012;14(3):281–292. doi: 10.31887/DCNS.2012.14.3/pchaste.
35. Maron E, Nutt D. Biological markers of generalized anxiety disorder. Dialogues Clin. Neurosci. 2017;19(2):147–158. doi: 10.31887/DCNS.2017.19.2/dnutt.
36. O'Connor PJ, Hill A, Kaya M, Martin B. The measurement of emotional intelligence: A critical review of the literature and recommendations for researchers and practitioners. Front. Psychol. 2019;10. doi: 10.3389/fpsyg.2019.01116
37. Shahzad S, Riaz Z, Begum N, Khanum SJ. Urdu translation and psychometric properties of Trait Emotional Intelligence Questionnaire Short Form (TEIQue-SF). Asian J. Manag. Sci. Educ. 2014;3(1):130–140. Available at www.leena-luna.co.jp.
38. Comfort A. The biology of senescence. J. Inst. Actuar. 1956;82(2):276–277. Available at https://www.cambridge.org/core/product/identifier/S0020268100046539/type/journal_article
39. Ferrucci L, Gonzalez-Freire M, Fabbri E, Simonsick E, Tanaka T, Moore Z, et al. Measuring biological aging in humans: A quest. Aging Cell 2020;19(2):2–3. doi: 10.1111/acel.13080
40. Ludwig FC, Smoke ME. The measurement of biological age. Exp. Aging Res. 1980;6(6):497–522. doi: 10.1080/03610738008258384.
41. Barak B, Gould S. Alternative age measures: A research agenda. Adv. Consum. Res. 1985;12:53. Available at https://www.acrwebsite.org/volumes/6357/volumes/v12/NA-12
42. Cao X, Ma C, Zheng Z, He L, Hao M, Chen X, et al. Contribution of life course circumstances to the acceleration of phenotypic and functional aging: A retrospective study. eClinicalMedicine. 2022;51:101548. doi: 10.1016/j.eclinm.2022.101548
43. K. Warner SL, Schaie W. Handbook of the Psychology of Aging. London: 2010.
44. Peters R. Ageing and the brain. Postgrad. Med. J. 2006;82(964):84–88. doi: 10.1136/pgmj.2005.036665.

45. Nichols H. What happens to the brain as we age? Medical News Today, 2017. Available at https://www.medicalnewstoday.com/articles/319185.php (accessed December 3, 2021).

46. Dercon Q, Nicholas JM, James SN, Schott JM, Richards M. Grip strength from midlife as an indicator of later-life brain health and cognition: Evidence from a British birth cohort. BMC Geriatr. 2021;21(1):1–11. doi: 10.1186/s12877-021-02411-7.

47. Cattaneo G, Bartrés-Faz D, Morris TP, Sánchez JS, Macià D, Tarrero C, et al. The Barcelona brain health initiative: A cohort study to define and promote determinants of brain health. Front. Aging Neurosci. 2018;10. doi: 10.3389/fnagi.2018.00321

48. Pauwels L, Chalavi S, Swinnen SP. Aging and brain plasticity. Aging (Albany NY). 2018;10(8):1789–1790. doi: 10.18632/aging.101514.

49. Gorlé N, Van Cauwenberghe C, Libert C, Vandenbroucke RE. The effect of aging on brain barriers and the consequences for Alzheimer's disease development. Mamm. Genome 2016; 27(7–8):407–420. doi: 10.1007/s00335-016-9637-8.

50. Zetterberg H, Blennow K. Chronic traumatic encephalopathy: Fluid biomarkers. In: Handbook of Clinical Neurology (1st ed., pp. 323–33). Elsevier B.V., 2018. doi: 10.1016/B978-0-444-63954-7.00030-6

51. Tandon R. What is a biomarker? Definitions, types, and research applications explained. 2016. Available at https://bitesizebio.com/26559/biomarkers-explained/ (accessed July 25, 2022).

52. Shaw G. The use and potential of pNF-H as a general blood biomarker of axonal loss: An immediate application for CNS injury. Brain Neurotrauma: Molecular, Neuropsychological, and Rehabilitation Aspects. Florida: CRC Press/Taylor & Francis, 2015. Available at http://www.ncbi.nlm.nih.gov/pubmed/26269910.

Chapter 5

Measuring Brain Health: Methodology and Insights

Introduction

The puzzle of how to measure brain health resembles the challenge of solving a Rubik's cube. How do we really know the state of this organ, which shapes our thoughts, actions, emotions, and decisions? In this chapter, we go through the detailed methodology of the research project that greatly increased understanding of brain health measurement. The study was done in four phases where each contributed significantly toward the overall picture of understanding the health of our brains (Figure 5.1).

Phase 1: Laying the Foundation

The foundation for our study was laid by broadly reviewing the key faculties of the brain from scientific literature, specifically deep diving into scientific records (Figure 5.2).

DOI: 10.1201/9781032676579-5

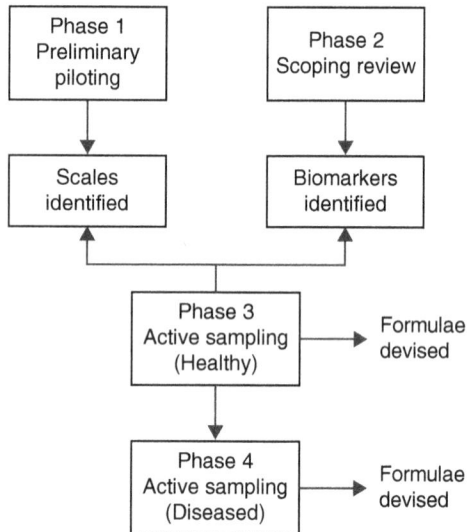

Figure 5.1   Flowchart showing general methodology followed in this research project.

To achieve this aim, we sought consultation with specialists, including neurologists, ophthalmologists, ENT specialists, and psycho-physiologists regarding which measures would best capture brain functions when used in real-life clinical settings. Consider this as brainstorming session with some great minds in medical care. During this meeting, these experts shared

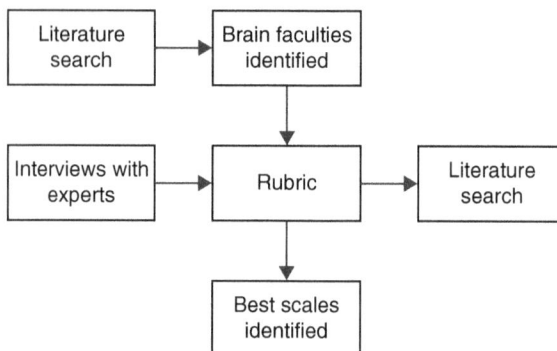

Figure 5.2   Flowchart showing methodology of Phase 1.

their experiences and insights guiding us to choose tools like visual field tests for vision, pure tone audiometry for hearing, hand grip dynamometry for motor strength, and the Mini-Mental State Examination (MMSE) for cognition.

This foundation is crucial because it enables us to understand different faculties possessed by the brain and how they could help in discovering human potentials. It reminds one of the historical anecdote recounted by Thomas Edison who was perceived as a slow learner during his early ages, not to mention undergoing dismissal from school. But this didn't stop him from reaching such heights in terms of innovation that have changed modern-day living standards, making him one of the most prolific inventors ever. His story exemplifies how understanding different brain faculties, from cognitive processing to creativity, can reveal unique pathways to unlocking human potential.

By the conclusion of Phase 1, a powerful toolkit was prepared so as to take on this challenge of evaluating brain health much like a canon of tools that depend upon the functions of the brain to determine its health—an analogy to what Edison did with his faculties to overcome initial learning difficulties and succeed.

Phase 2: Biomarker Scavenger Hunt

Phase 2 involved identifying potential biomarkers for brain injury (Figure 5.3). It was more or less a biological scavenger hunt during which medical literature was combed through dating back to 1930s when "punch drunk" was first used to refer to repetitive head injuries. Blood proteins that were indicative of brain damage were searched—biomarkers that indicated if the brain was distressed.

At this point, some proteins like neurofilament light (NfL), neuron-specific enolase (NSE), and phosphorylated tau (pTau)

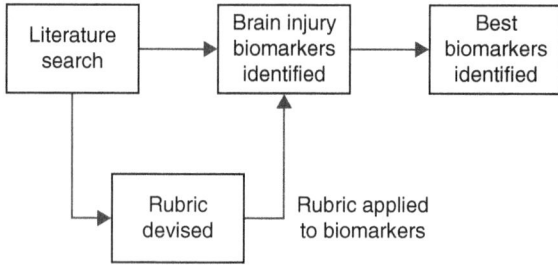

Figure 5.3   Flowchart showing methodology of Phase 2.

were investigated through the use of databases such as Google Scholar and MEDLINE. Those proteins answered different questions concerning neuron health just as certain evidences in a criminal scene are peculiar to detectives, hence aiding in knowing what happened where and how it was done. Therefore, a scoring system was devised that considered factors like specificity and sensitivity—essentially, how accurately these proteins could detect brain injury.

This can be equated with Alois's Alzheimer's work, who, in 1906 observed the characteristic plaques and tangles associated with Alzheimer's disease. This work is fundamental in setting a stage for further research across generations as to how there could be more accurate brain injuries/degenerations prediction measures.

Phase 3: Testing the Formula on Healthy Participants

Phase 3 augmented the theory developed in Phase 1, as it constituted a cross-sectional study with 75 healthy individuals using the tools from Phase 1 and biomarkers from Phase 2 (Figure 5.4). At this point, rubber met the road. Five aspects were tested for each person, including visual fields, hearing abilities, cognition levels (depression symptoms were also measured), emotional well-being, and hand grip strength. Additionally, blood samples were collected from each subject for measuring

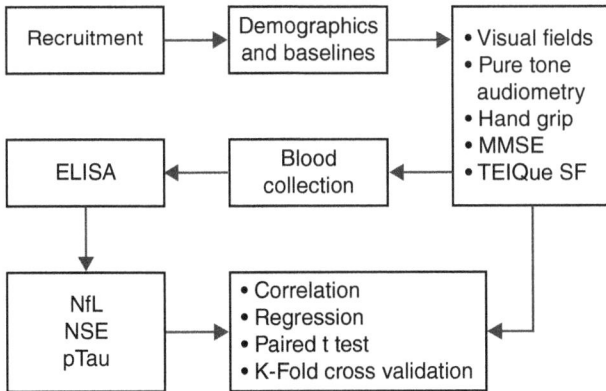

Figure 5.4   Flowchart showing methodology of Phase 3.

pTau, NSE, and NfL concentrations in blood. In essence, this was decoding health patterns related to brain using data obtained from tests themselves; this was similar to reverse-engineering the brain's health code—seeing if the data gathered could predict what was happening on a molecular level.

For instance, the Framingham Heart Study transformed cardiology practice by systematically measuring cardiovascular health risk factors in large cohorts. In the same way, Phase 3 aimed to systematically evaluate the predictors of brain health in a healthy population. Multiple linear regression models that predict biomarker levels were created based on given test values. Comparing predicted levels against observed levels provided clues regarding which methods are more reliable when evaluating brain health.

Phase 4: Putting the Formula to the Test in Disease

Following positive outcomes from healthy participants in Phase 3, Phase 4 was commenced. Here the same recipe was applied to those suffering from neurological disorders. The groups here were those struggling with amblyopia,

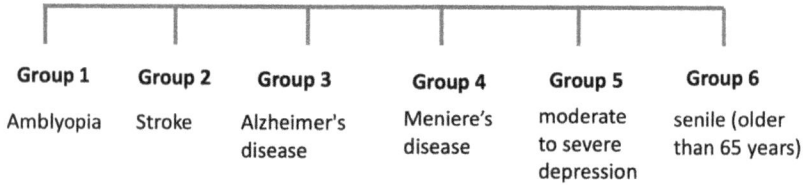

Group 1	Group 2	Group 3	Group 4	Group 5	Group 6
Amblyopia	Stroke	Alzheimer's disease	Meniere's disease	moderate to severe depression	senile (older than 65 years)

Figure 5.5   Inclusion criteria.

stroke, Alzheimer's, Meniere disease, moderate or severe depression, and age-related decline (Figure 5.5). This diverse group allowed us to test whether the predictive formulas developed in Phase 3 worked equally well for individuals with neurological conditions.

Would an Alzheimer patient's markers and measurements be the same as those of a 25-year-old without any health challenges? This phase not only confirmed its reliability but also unearthed some new findings. For instance, while the core principles held true, each disease introduced unique variables that required slight modifications. Akin to early treatments of Parkinson's disease introduced by Jean-Martin Charcot, this also required adjustments depending on individual patient symptoms. Our findings in Phase 4 underscored the need for tailored approaches, emphasizing the adaptability required in clinical practice (Figure 5.6).

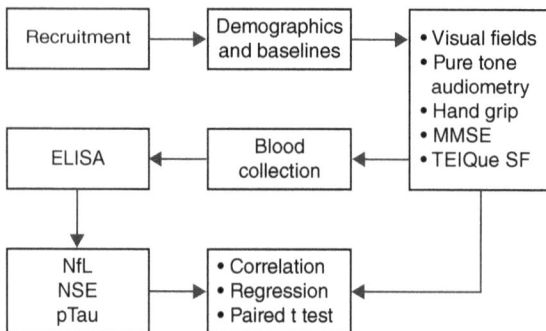

Figure 5.6   Flowchart showing methodology of Phase 4.

The Grandeur of It All: Mixed Methods and Validation

This is a study which used mixed-method strategy by incorporating qualitative insights derived from some interviews with qualitative and quantitative scrutiny of data obtained from empirical analysis process. Doctors always use hands-on bedside approach when diagnosing sick patients, hence the use of interviews with experts who have used systematic empirical analysis method gave detailed insight that empowered us to know brain health determinants in totality.

We also conducted strict cross-validation by segmenting the data and then testing different models frequently to see if the results change or not. This was done through an extensive process that helped to establish the consistency in the findings rather than seeing them as just being a myth. Through the cross-validation process, the techniques of measuring brain health were strengthened which enhanced their reliability to a greater degree. Just as the famous Stanford Marshmallow Experiment highlighted the importance of delayed gratification and validated its conclusions over time, our cross-validation efforts reinforced the reliability of our methods for measuring brain health.

A Logical Framework for Brain Health

It is quite complex to comprehend brain health, but this chapter has covered some ground in order to arrive at the right model with regard to its four stages of development. From identifying critical brain functions and selecting biomarkers to devising predictive formulas and validating these in both healthy and diseased populations, we brought the science of brain health closer to everyday clinical practice.

In future, we believe these methods will help doctors detect signs of cognitive decline much earlier than current diagnostic criteria allow, hence providing early intervention options leading to better outcomes. Much like the Framingham Study transformed cardiovascular care, this research aims to offer a structured, evidence-based approach to understanding and nurturing brain health. By embracing both quantitative measures and the lived experiences of individuals, we move toward a future where brain health is understood and addressed with the same precision and compassion as any other aspect of medical care.

Bibliography

1. Poldrack RA, Huckins G, Varoquaux G. Establishment of best practices for evidence for prediction. JAMA Psychiatry. 2020;77(5):534. doi: 10.1001/jamapsychiatry.2019.3671.
2. Harrell FE, Lee KL, Califf RM, Pryor DB, Rosati RA. Regression modelling strategies for improved prognostic prediction. Stat. Med. 1984;3(2):143–152. doi: 10.1002/sim.4780030207.
3. Miché M, Studerus E, Meyer AH, Gloster AT, Beesdo-Baum K, Wittchen HU, et al. Prospective prediction of suicide attempts in community adolescents and young adults, using regression methods and machine learning. J. Affect. Disord. 2020;265: 570–578. doi: 10.1016/j.jad.2019.11.093.
4. Bobbitt Z. A quick intro to leave-one-out cross-validation (LOOCV). Statology, 2023.
5. Charmaz K. Qualitative Research: Grounded Theory. SAGE Publications. 2003. Available at https://guides.temple.edu/groundedtheory (accessed August 7, 2022).
6. O'Brien BC, Harris IB, Beckman TJ, Reed DA, Cook DA. Standards for reporting qualitative research: A synthesis of recommendations. Acad. Med. 2014;89(9):1245–1251. doi: 10.1097/ACM.0000000000000388.
7. Sandelius Å, Zetterberg H, Blennow K, Adiutori R, Malaspina A, Laura M, et al. Plasma neurofilament light chain concentration in the inherited peripheral neuropathies. Neurology. 2018;90(6):e518–e524. doi: 10.1212/WNL.0000000000004932.

8. Stammet P, Collignon O, Hassager C, Wise MP, Hovdenes J, Åneman A, et al. Neuron-specific enolase as a predictor of death or poor neurological outcome after out-of-hospital cardiac arrest and targeted temperature management at 33°C and 36°C. J. Am. Coll. Cardiol. 2015;65(19):2104–2114. doi: 10.1016/j.jacc.2015.03.538.

9. Delmotte K, Schaeverbeke J, Poesen K, Vandenberghe R. Prognostic value of amyloid/tau/neurodegeneration (ATN) classification based on diagnostic cerebrospinal fluid samples for Alzheimer's disease. Alzheimers Res Ther. 2021;13(1):84. doi: 10.1186/s13195-021-00817-4.

10. Meniere's disease. Mayo clinic. 2022. Available at https://www.mayoclinic.org/diseases-conditions/menieres-disease/diagnosis-treatment/drc-20374916 (accessed July 28, 2022).

11. Diagnostic criteria of stroke. p. 1–5. Available at https://www.cuore.iss.it/eurociss/reg_ictus/pdf/ictus_criteri-diagnostici.pdf (accessed June 20, 2022).

12. Williams C. Amblyopia. BMJ Clin. Evid. 2009;2009:1–17. Available at http://www.ncbi.nlm.nih.gov/pubmed/21726480

13. Alzheimer' s Disease Diagnostic Guidelines. National Institute on Aging-Alzheimer's Association, 2019, p. 2–5. Available at https://www.nia.nih.gov/health/alzheimers-disease-diagnostic-guidelines (accessed April 28, 2022).

14. Truschel J. Depression definition and DSM-5 diagnostic criteria. PSYCOM, 2019. Available at https://www.psycom.net/depression-definition-dsm-5-diagnostic-criteria/

15. Tolentino JC, Schmidt SL. DSM-5 criteria and depression severity: Implications for clinical practice. Front. Psychiatry 2018;9:1–9.

16. Lee SB, Oh JH, Park JH, Choi SP, Wee JH. Differences in youngest-old, middle-old, and oldest-old patients who visit the emergency department. Clin. Exp. Emerg. Med. 2018;5(4): 249–255. doi: 10.15441/ceem.17.261.

17. Casarella J. The brain and mental illness. J. Neurol. Neurosurg. Psychiatry 2021;11(44):342–345.

18. Kumar Seetlani N, Kumar N, Mubeen KI, Ali A, Shams N, Sheikh T. Alzheimer and vascular dementia in the elderly patients. Pakistan J. Med. Sci. 2016;32(5):1286–1290. doi: 10.12669/pjms.325.10792.

19. Cooper A, Petrides KV. A psychometric analysis of the Trait Emotional Intelligence Questionnaire-Short Form (TEIQue-SF) using item response theory. J. Pers. Assess. 2010;92(5):449–457. doi: 10.1080/00223891.2010.497426.

20. Akinkugbe OO. Role of teaching hospitals in a developing country. BMJ. 1973;1(5851):474–476. doi: 10.1136/bmj.1.5851.474.

21. Hussain R, Ali W, Sohaib M. Perceptions of physicians regarding implementation of hospital management information systems in a tertiary setting hospital of a developing country. Cureus. 2021;13(10):e18674. doi: 10.7759/cureus.18674.

22. Hamm LM, Anstice NS, Black JM, Dakin SC. Recognition acuity in children measured using the Auckland optotypes. Ophthalmic Physiol. Opt. 2018;38(6):596–608. doi: 10.1111/opo.12590.

23. Chou R, Dana T, Bougatsos BC. Screening for visual impairment in children ages 1–5 years: Systematic review to update the 2004 U.S. Preventive Services Task Force Recommendation. In: U.S. Preventive Services Task Force Evidence Syntheses, Formerly Systematic Evidence Reviews. Rockville (MD): Agency for Healthcare Research and Quality, 2011. Available at https://www.ncbi.nlm.nih.gov/books/NBK52708/.

24. Nakai Y, Bessho K, Shono Y, Taoka K, Nakai Y. Comparison of imo and Humphrey field analyzer perimeters in glaucomatous eyes. Int. J. Ophthalmol. 2021;14(12):1882–1887. doi: 10.18240/ijo.2021.12.11.

25. Talbot R, Goldberg I, Kelly P. Evaluating the accuracy of the visual field index for the Humphrey visual field analyzer in patients with mild to moderate glaucoma. Am. J. Ophthalmol. 2013;156(6):1272–1276. doi: 10.1016/j.ajo.2013.07.025.

26. Vermiglio AJ, Soli SD, Fang X. An argument for self-report as a reference standard in audiology. J. Am. Acad. Audiol. 2018;29(3):206–222. doi: 10.3766/jaaa.16128.

27. Swanepoel DW, Mngemane S, Molemong S, Mkwanazi H, Tutshini S. Hearing assessment: Reliability, accuracy, and efficiency of automated audiometry. Telemed. J. E Health 2010;16(5):557–563. doi: 10.1089/tmj.2009.0143.

28. Biagio-de Jager L, van Dyk Z, Vinck BH. Diagnostic accuracy of CE Chirp. Int. J. Pediatr. Otorhinolaryngol. 2020;135:110071. doi: 10.1016/j.ijporl.2020.110071.

29. Bhatt J, Kuchhal V, Saklani K, Kumar V. Accuracy of OAE and BERA to detect the incidence of hearing loss in newborn. J. Evol. Med. Dent. Sci. 2015;4(49):8466–8474. doi: 10.14260/jemds/2015/1228.

30. Erdodi LA, Abeare CA, Lichtenstein JD, Tyson BT, Kucharski B, Zuccato BG, et al. Wechsler adult intelligence scale-fourth edition (WAIS-IV) processing speed scores as measures of noncredible responding: The third generation of embedded performance validity indicators. Psychol. Assess. 2017;29(2): 148–157. doi: 0.1037/pas0000319.

31. Zarantonello MM, Munley PH, Milanovich J. Predicting Wechsler adult intelligence scale-revised (WAIS-R) IQ scores from the Luria-Nebraska neuropsychological battery (form I). J. Clin. Psychol. 1993;49(2):225–233. doi: 10.1002/1097-4679(199303)49:2%3C225::AID-JCLP2270490216%3E3.0.CO;2-C.

32. Pezzotti P, Scalmana S, Mastromattei A, Di Lallo D. The accuracy of the MMSE in detecting cognitive impairment when administered by general practitioners: A prospective observational study. BMC Fam Pract. 2008;9(1):11. doi: 10.1186/1471-2296-9-29.

33. Spering CC, Hobson V, Lucas JA, Menon CV, Hall JR, O'Bryant SE. Diagnostic accuracy of the MMSE in detecting probable and possible Alzheimer's disease in ethnically diverse highly educated individuals: An analysis of the NACC database. J. Gerontol. A Biol. Sci. Med. Sci. 2012;67(8):890–896. doi: 10.1093/gerona/gls006.

34. Louis TM, Barchard KA. Examining the reliability and validity of the MSCEIT faces scale. 2005, p. 1–2. Available at https://img.faculty.unlv.edu/lab/conference-presentations/conference-posters/MSCEITFaces.pdf

35. Gafner SC, Bastiaenen CH, Ferrari S, Gold G, Terrier P, Hilfiker R, et al. Hip muscle and hand-grip strength to differentiate between older fallers and non-fallers: A cross-sectional validity study. Clin. Interv. Aging 2018;13:1–8. doi: 10.2147/CIA.S146834.

36. Amaral JF, Mancini M, Novo Júnior JM. Comparison of three hand dynamometers in relation to the accuracy and precision of the measurements. Braz. J. Phys. Ther. 2012;16(3):216–224. doi: 10.1590/s1413-35552012000300007.

37. Cuthbert SC, Goodheart GJ. On the reliability and validity of manual muscle testing: A literature review. Chiropr. Osteopat. 2007;15(1):4. doi: 10.1186/1746-1340-15-4.

38. Gallamini M, Piastra G, Lucarini S, Porzio D, Ronchi M, Pirino A, et al. Revisiting the instrumented Romberg test: Can today's technology offer a risk-of-fall screening device for senior citizens? An experience-based approach. Life. 2021;11(2):161. doi: 10.3390/life11020161.

39. Honaker JA, Boismier TE, Shepard NP, Shepard NT. Fukuda stepping test: Sensitivity and specificity. J. Am. Acad. Audiol. 2009;20(5):311–314. doi: 10.3766/jaaa.20.5.4.

40. Santo AL, Reed JM, Lynall RC. Tandem gait test performance in healthy, physically active adults: Clinical implications for concussion evaluation. J. Sci. Med. Sport 2021;24(7):622–626. doi: 10.1016/j.jsams.2021.01.005.

Chapter 6

Laying the Basis for Measuring Brain Health

Introduction

Brain as an organ is unique—indeed, it ranks higher than any other organ in terms of prestige, commanding power, and conducting activities. It is like an orchestra conductor who knows how to make use of various instruments to produce melodious sounds. However, its activities are faster than one can think could ever be possible. Consider this: Every second that passes by, our brains takes in so much information from its surroundings as if it were receiving emails nonstop. But unlike our inbox, which we might ignore until it reaches breaking point, our brain would always respond whenever a piece of information arrives. It is like a supercomputer on steroids, outpacing the fastest tech we have today. Your latest smartphone? Just a glorified pocket calculator in comparison.

But how do we know if this supercomputer is still operating at its best? According to American Heart Association/ American Stroke Association (AHA/ASA), brain health can be

DOI: 10.1201/9781032676579-6

thought of as a school report card—remember those? An "A" means you're excelling, while anything less suggests there's room for improvement. The AHA/ASA defines brain health as the "average performance levels among people of that age who are free of known brain or major organ diseases." In other words, you want to be at or above this standard for your age—no cutting corners!

So how do we measure brain health in reality? This is where we will now need to deploy a variety of instruments and tests like detectives combing through a crime scene. Numerous tools provide glimpses into how our brains work, including motor function which refers basically to how well one moves around or maintains coordination. This function depends upon key areas such as the motor cortex, cerebellum, and basal ganglia that can be compared to the brain's dream team ensuring one is able to walk or chew gum without tripping over their own two feet.

If you are still able to think fast or handle numbers with ease, then it is highly probable that your prefrontal cortex—the top boss of your brain—is still in good shape. That's excellent news! Various tools are used to track these different brain areas, which range from structural anatomy (think of this as taking the ultimate brain selfie) to biochemical activity (imagine your brain like a laboratory with bubbling beakers in it).

Let us delve deeper into this matter. Brodmann meticulously mapped out the brain into 52 different regions, each of which has its own character like that of a neighborhood. Different neighborhoods might act like small towns where everything runs smoothly or they may cooperate and form coalitions just as the Avengers did when saving the planet.

But how do we tell what is happening in these cerebral neighborhoods? We have sophisticated tools. As far as cognitive assessment goes, there are tests such as Mini-Mental State Examination (MMSE) and Wechsler scale; essentially the spontaneous awakening trials (SATs) for your brain. Want to see how you stack up against other geniuses? These tests will help you know if your cognitive faculties are up to the mark. And for checking hearing abilities, pure tone audiometry (PTA) and brain-evoked response audiometry (BERA) ensure that both your ears and brain are still in sync.

Motor skills—such as balance, strength, and coordination—are vital for keeping you agile and in sync. Whether it is catching a ball, threading a needle, or dancing the Macarena without a misstep, we use tests that evaluate strength, endurance, and hand–eye coordination to determine how well your motor functions are holding up. If you can manage all these with ease, your motor control is in good shape!

However, measuring brain health remains an enigma, especially in the developing countries. Conducting these tests seems unreasonable like trying to conduct a symphony with missing instruments if anyone asks. The hurdles are numerous: high costs, limited equipment, scarcity of skilled personnel, and overwhelming hospital queues that require a saint's patience. Most people who need brain diagnostics cannot afford them at all due to poverty coupled with political issues.

This is where our research steps in, with a clear mission—to identify the most effective, efficient tools for assessing brain health in clinical settings, especially in developing countries. Think of it as a mission carried out by Sherlock Holmes—pipe in hand—trying to get hold of the simple methods of diagnosing brain disorders. This isn't just a clinical investigation; it is a full-fledged adventure to make brain health assessments accessible to all. After all, ensuring the brain's health is more than just a scientific endeavor; it is an essential pursuit for human well-being.

Playing Brain Detective—With a Magnifying Glass and a Notebook

We began our Brain Detective Work at the Institute of Basic Medical Sciences, Khyber Medical University, Pakistan, during early 2022. An investigation needs permissions and we got ours approved by the ethical board (even brain detectives have to follow the rules). Our approach was three-pronged— similar to a detective solving a high-profile case. We began by thoroughly reviewing medical literature to identify the brain's most essential functions, eventually narrowing it down to five key areas. It felt like selecting the right puzzle pieces from a jumble of options to crack the code. Following an extensive literature review, we conducted interviews with professionals—such as ophthalmologists and ENT specialists—who served as our trusted "Watsons," providing expert insights from their specific fields. These individuals were chosen for their deep knowledge and relevance to the research. To systematically evaluate each identified tool, we then created a rating scale—much like grading a science project or judging a competitive cooking show—to determine each tool's strengths and weaknesses.

The ophthalmologists, ENT (ear, nose, throat) specialists, psychologists, and neurologists, among others, were asked questions related to brain healthcare. Some of these questions pertaining to visual function were as follows:

1. Which is the best technique for measuring visual function based on its feasibility, most common usage, and ability to diagnose visual impairment?
2. Do you think optimum visual function has a link with brain health? If Yes, Why?

Everything was documented (with permission from the source because even brain investigators should respect individual privacy) and detailed field notes were taken so as to have all pieces of information. This information was then analyzed using grounded theory—an approach that involves sorting through information for patterns, clues, and ultimately solving a case. Also, since no detective is ubiquitous, everything was meticulously cross-checked by the Principal Investigator in order to ensure reliability, validity, as well as accuracy (Table 6.1).

The Best Tools in Our Brainy Toolbox— Like a Detective's Arsenal

During an extensive period of inquiry and investigation on brain functions necessary for good health, five brain functions were identified—visual function, hearing, cognition, motor function, and emotional well-being. Precise measuring instruments were needed for each one.

Visual Function

Visual fields testing dominated other methods such as visual acuity as far as visual function is concerned. The eyes would behave like a top of the line wide-angle lens camera, if there is synchronization between what the brain sees and what your eyes report to it. Visual acuity, while still important, just can't compete with the broader perspective that visual fields offer.

Hearing Ability

In hearing, PTA seems to be the choice that beats all others. It could be likened to the Swiss army knife of hearing tests, cost-effective, widely used, and reliable. It is always there when you need it; that is why, it was our best option.

Table 6.1 Rubric with Grading for Choosing the Appropriate Scales

Criteria S. No.	Criteria	Rubric for the Scales			
		Poor Indicator 1 Point	Average Indicator 2 Points	Good Indicator 3 Points	
Criterion 1	Price per test for patient (dollars)	More than 5 dollars	1–5 dollars	Free	
Criterion 2	Portability	Completely immobile to one place	Can be moved within an institute	Fully portable	
Criterion 3	Level of skill needed for use	Proper qualification required	Training needed	No prior training	
Criterion 4	Assessment time	More than 15 minutes	6–15 minutes	1–5 minutes	
Criterion 5	AUC	$0.6 < AUC < 0.8$	$0.8 < AUC < 0.9$	$0.9 < AUC < 1.0$	
Criterion 6	Diagnosing accuracy/ validity/ reliability	Less than 69%	70–89%	90–100%	
Criterion 7	Usage in clinical setup/ research	Usage less than 49%	Usage between 50% and 79%	Usage more than 80%	
Criterion 8	Link with brain health	Less than 5	6–8	9–10	

Cognition

Here, the MMSE scale came first due to some valid reasons. It is short, simple, and free—you won't be looking for anything more than that in terms of cognitive tests. Indeed, assessment of cognitive function using MMSE is regarded with high-esteem because it is famous among many scholars as well as serves as the final verdict for cognition testing. If your MMSE scores are solid, your brain is likely firing on all cylinders.

Motor Function

Muscle strength was assessed using manual muscle testing (MMT). To assess muscle strength, MMT was the preferred choice. It feels like an old friend always ready to lend a hand—just a simple muscle check that doesn't require any pricey equipment. This is user-friendly and helpful in clinical settings where resources are scarce or hard to come by. But its shortcoming is its subjectivity, i.e., it depends upon the human experience. The findings attained will vary depending on whether an experienced or a novice physician performs the MMT. This glitch can be overcome by a more objective tool—the dynamometer, which measures muscle strength in kilograms.

Emotional Well-Being

Our inventory of choice for evaluating emotional health was TEIQue-SF scale. It is simple and quick, and has a broader utility (Table 6.2).

Why These Tools? Unpacking the Detective's Notebook

Choosing the best tool for each brain function was like trying to pick the ultimate superhero team—each tool has

Table 6.2 Scoring of Tools According to the Rubric Devised

S. No.	Brain Functions	Tools	C1: Price Per Test for Patient	C2: Portability	C3: Level of Skill Needed for Use	C4: Time for Assessment	C5: AUC	C6: Diagnosing Accuracy	C7: Usage in Clinical Setup	C8: Link with Brain Health	Total
1.	Visual function	Visual acuity	2	1	2	1	3[145]	3[146]	3	2	**17**
2.		Visual fields	3	2	2	2	2[147]	2[148]	3	1	**17**
3.	Hearing function	Pure tone audiometry	2	2	2	2	1[149]	3[150]	3	1	**16**
4.		Brain-evoked response audiometry	1	2	2	2	1[151]	3[152]	1	1	13
5.	Cognition	Wechsler's scale	1	3	1	1	2[153]	1[154]	1	3	13
6.		MMSE scale	3	3	3	2	2[155]	2[156]	2	3	**20**
7.	Emotional well being	TEIQue-SF	3	3	3	2	2	2[142]	1	2	**18**
8.		MSCEIT	1	3	1	1	2	1[157]	1	2	12
9.	Motor function	Hand grip strength Dynamometry	3	3	3	1[158]	2[159]	1	2	**18**	**18**
10.		Manual muscle testing	3	2	3	1	1[160]	3	2	18	18
11.		Balance Romberg's test	3	3	3	1[161]	1	3	2	**19**	**19**
12.		Unterberger's test	3	3	3	1[162]	1	2	2	18	18
13.		Tandem walking test	3	3	3	1[163]	1	2	2	18	18

C1–C8: Criterion 1–Criterion 8, MMSE: Mini-Mental State Examination, TEIQue-SF: Trait Emotional Intelligence Questionnaire Short Form, MSCEIT: Mayer–Salovey–Caruso emotional intelligence test.

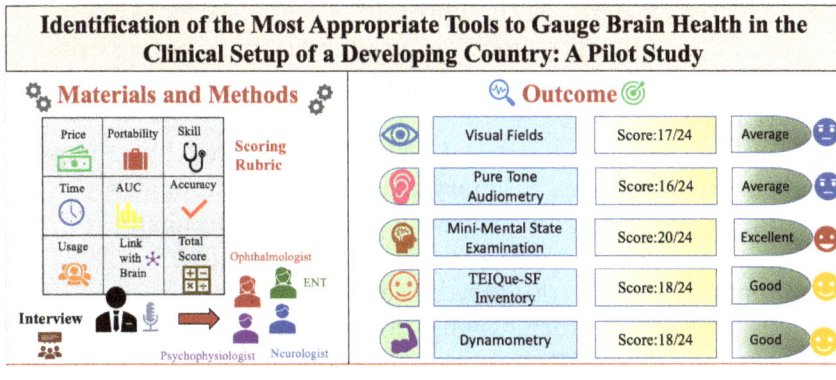

Figure 6.1 Methodology followed and outcomes of Phase 1 of the research conducted.

its own strengths, but only a few truly save the day. We made our selections based on a range of factors, including price, portability, and accuracy (Figure 6.1). For instance, visual fields surpass visual acuity because of its stronger link to brain function and better accuracy. It is like choosing Superman over Batman—both are great, but one has an extra edge when it comes to saving the world (or, in this case, assessing brain health).

In the end, we didn't just choose these tools because they're effective; we chose them because they're practical, especially in the context of a developing country.

Bibliography

1. Casarella J. The brain and mental illness. J. Neurol. Neurosurg. Psychiatry. 2021;11(44):342–345.
2. Wang Y, Pan Y, Li H. What is brain health and why is it important? BMJ 2020;371. doi: 10.1136/bmj.m3683.
3. Guyton AC, Hall JE. Textbook of Medical Physiology (13th ed.). Elsevier Inc., 2016.
4. Lo RY. Longitudinal change of biomarkers in cognitive decline. Arch. Neurol. 2011;68(10):1257. doi: 10.1001/archneurol.2011.123.

5. Boyle PA, Wang T, Yu L, Wilson RS, Dawe R, Arfanakis K, et al. The "cognitive clock": A novel indicator of brain health. Alzheimers Dement. 2021; 17(12):1923–1937. doi: 10.1002/alz.12351.

6. Ferng A. Brodmann Areas. KENHUB. 2021. Available at https://www.kenhub.com/en/library/anatomy/brodmann-areas.

7. Lloyd D. Scanning the neurocracy: What do Brodmann areas do? Lloydia 1987;11:23–78. Available at http://commons.trincoll.edu/dlloyd/research/videos-web-pages/scanning-the-neurocracy-what-do-brodmann-areas-do/

8. Mini-Mental State Examination. Physiopedia, 2021. Available at https://www.physio-pedia.com/Mini-Mental_State_Examination (accessed December 3, 2021).

9. Kumar Seetlani N, Kumar N, Mubeen KI, Ali A, Shams N, Sheikh T. Alzheimer and vascular dementia in the elderly patients. Pak. J. Med. Sci. 2016;32(5):1286–1290. Available at http://pjms.com.pk/index.php/pjms/article/view/10792

10. Walker JJ, Cleveland LM, Davis JL, Seales JS. Audiometry screening and interpretation. Am. Fam. Physician 2013;87(1):41–47.

11. Rovner BW, Casten RJ, Massof RW, Leiby BE, Tasman WS. Psychological and cognitive determinants of vision function in age-related macular degeneration. Arch. Ophthalmol. 2011;129(7):885–890. doi: 10.1001/archophthalmol.2011.146.

12. Reuben DB, Magasi S, McCreath HE, Bohannon RW, Wang YC, Bubela DJ, et al. Motor assessment using the NIH Toolbox. Neurology 2013;80(Suppl. 3):65–75. doi: 10.1212/WNL.0b013e3182872e01

13. Akinkugbe OO. Role of teaching hospitals in a developing country. BMJ 1973;1(5851):474–476. doi: 10.1136/bmj.1.5851.474.

14. Hussain R, Ali W, Sohaib M. Perceptions of physicians regarding implementation of hospital management information systems in a tertiary setting hospital of a developing country. Cureus 2021;13(10):e18674. doi: 10.7759/cureus.18674.

15. Charmaz K. Qualitative Research: Grounded Theory. SAGE Publications, 2003. Available at https://guides.temple.edu/groundedtheory.

16. Hamm LM, Anstice NS, Black JM, Dakin SC. Recognition acuity in children measured using the Auckland optotypes. Ophthalmic. Physiol. Opt. 2018;38(6):596–608. doi: 10.1111/opo.12590.

17. Chou R, Dana T, Bougatsos BC. Screening for Visual Impairment in Children Ages 1–5 Years: Systematic Review to Update the 2004 U.S. Preventive Services Task Force Recommendation. In: U.S. Preventive Services Task Force Evidence Syntheses, Formerly Systematic Evidence Reviews. Rockville, MD: Agency for Healthcare Research and Quality, 2011. Available at https://www.ncbi.nlm.nih.gov/books/NBK52708/

18. Nakai Y, Bessho K, Shono Y, Taoka K, Nakai Y. Comparison of imo and Humphrey field analyzer perimeters in glaucomatous eyes. Int. J. Ophthalmol. 2021;14(12):1882–1887. doi: 10.18240/ijo.2021.12.11.

19. Talbot R, Goldberg I, Kelly P. Evaluating the accuracy of the visual field index for the Humphrey visual field analyzer in patients with mild to moderate glaucoma. Am. J. Ophthalmol. 2013;156(6):1272–1276. doi: 10.1016/j.ajo.2013.07.025.

20. Vermiglio AJ, Soli SD, Fang X. An argument for self-report as a reference standard in audiology. J. Am. Acad. Audiol. 2018;29(3):206–222. doi: 10.3766/jaaa.16128.

21. Swanepoel DW, Mngemane S, Molemong S, Mkwanazi H, Tutshini S. Hearing assessment: Reliability, accuracy, and efficiency of automated audiometry. Telemed. J. E Health 2010;16(5):557–563. doi: 10.1089/tmj.2009.0143.

22. Biagio-de Jager L, van Dyk Z, Vinck BH. Diagnostic accuracy of CE Chirp. Int. J. Pediatr. Otorhinolaryngol. 2020;135:110071. doi: 10.1016/j.ijporl.2020.110071.

23. Bhatt J, Kuchhal V, Saklani K, Kumar V. Accuracy of OAE and BERA to detect the incidence of hearing loss in newborn. J. Evol. Med. Dent. Sci. 2015;4(49):8466–8474.

24. Erdodi LA, Abeare CA, Lichtenstein JD, Tyson BT, Kucharski B, Zuccato BG, et al. Wechsler adult intelligence scale-fourth edition (WAIS-IV) processing speed scores as measures of noncredible responding: The third generation of embedded performance validity indicators. Psychol. Assess. 2017;29(2):148–157.

25. Zarantonello MM, Munley PH, Milanovich J. Predicting Wechsler adult intelligence scale-revised (WAIS-R) IQ scores from the Luria-Nebraska neuropsychological battery (form I). J. Clin. Psychol. 1993;49(2):225–233. doi: 10.1002/1097-4679(199303)49:2<225::AID-JCLP2270490216>3.0.CO;2-C.

26. Pezzotti P, Scalmana S, Mastromattei A, Di Lallo D. The accuracy of the MMSE in detecting cognitive impairment when administered by general practitioners: A prospective observational study. BMC Fam. Pract. 2008;9:1–11. doi: 10.1186/1471-2296-9-29.

27. Spering CC, Hobson V, Lucas JA, Menon CV, Hall JR, O'Bryant SE. Diagnostic accuracy of the MMSE in detecting probable and possible Alzheimer's disease in ethnically diverse highly educated individuals: An analysis of the NACC database. J. Gerontol. A Biol. Sci. Med. Sci. 2012;67(8):890–896. doi: 10.1093/gerona/gls006.

28. Cooper A, Petrides KV. A psychometric analysis of the Trait Emotional Intelligence Questionnaire-Short Form (TEIQue-SF) using item response theory. J. Pers. Assess. 2010;92(5):449–457. doi: 10.1080/00223891.2010.497426.

29. Louis TM, Barchard KA. Examining the reliability and validity of the MSCEIT faces scale. 2005, p. 1–2. Available at https://img.faculty.unlv.edu/lab/conference-presentations/conferenceposters/MSCEITFaces.pdf

30. Gafner SC, Bastiaenen CH, Ferrari S, Gold G, Terrier P, Hilfiker R, et al. Hip muscle and hand-grip strength to differentiate between older fallers and non-fallers: A cross-sectional validity study. Clin. Interv. Aging 2018;13:1–8. doi: 10.2147/CIA.S146834.

31. Amaral JF, Mancini M, Novo Júnior JM. Comparison of three hand dynamometers in relation to the accuracy and precision of the measurements. Braz. J. Phys. Ther. 2012;16(3):216–224. doi: 10.1590/S1413-35552012000300007.

32. Cuthbert SC, Goodheart GJ. On the reliability and validity of manual muscle testing: A literature review. Chiropr. Osteopat. 2007;15(1):4. doi: 10.1186/1746-1340-15-4.

33. Gallamini M, Piastra G, Lucarini S, Porzio D, Ronchi M, Pirino A, et al. Revisiting the instrumented Romberg test: Can today's technology offer a risk-of-fall screening device for senior citizens? An experience-based approach. Life. 2021;11(2):161. doi: 10.3390/life11020161

34. Honaker JA, Boismier TE, Shepard NP, Shepard NT. Fukuda stepping test: Sensitivity and specificity. J. Am. Acad. Audiol. 2009;20(5):311–314. doi: 10.3766/jaaa.20.5.4.

35. Santo AL, Reed JM, Lynall RC. Tandem gait test performance in healthy, physically active adults: Clinical implications for concussion evaluation. J. Sci. Med. Sport 2021;24(7):622–626. doi: 10.1016/j.jsams.2021.01.005.

36. Magone MT, Kwon E, Shin SY. Chronic visual dysfunction after blast-induced mild traumatic brain injury. J. Rehabil. Res. Dev. 2014;51(1):71–80. doi: 10.1682/JRRD.2013.01.0008.

37. Armstrong RA. Visual problems associated with traumatic brain injury. Clin. Exp. Optom. 2018;101(6):716–726. doi: 10.1111/cxo.12670.

38. Tuwaig M, Savard M, Jutras B, Poirier J, Collins DL, Rosa-Neto P, et al. Deficit in central auditory processing as a biomarker of pre-clinical Alzheimer's disease. J. Alzheimer's Dis. 2017;60(4):1589–1600. doi: 10.3233/JAD-170545

39. Parker T, Cash DM, Lane C, Lu K, Malone IB, Nicholas JM, et al. Pure tone audiometry and cerebral pathology in healthy older adults. J. Neurol. Neurosurg. Psychiatry. 2020;91(2):172–176. doi: 10.1136/jnnp-2019-321897

40. Duc NT, Ryu S, Qureshi MNI, Choi M, Lee KH, Lee B. 3D-deep learning based automatic diagnosis of Alzheimer's disease with joint MMSE prediction using resting-state fMRI. Neuroinformatics 2020;18(1):71–86. doi: 10.1007/s12021-019-09419-w.

41. Dhillon NK, Linaval NT, O'Rourke J, Barmparas G, Yang A, Cho N, et al. How repetitive traumatic injury alters long-term brain function. J. Trauma Acute Care Surg. 2020;89(5):955–961. doi: 10.1097/TA.0000000000002811

Chapter 7

Digital Investigation of Traumatic Head Injuries

Introduction

Anyone can incur an accident, be it a rough football game or simply underestimating the height of your kitchen cabinet. In case the brain knocks against the skull, it leads to a small car accident inside one's head. Now imagine this happening repeatedly; it is not just a mere headache but also the onset of a more sinister thing called tauopathy. This can be metaphorically compared to a pair of unruly thugs (protein) being led by their notorious leader (tau) crashing into one's home (brain) and destroying it. This buildup can lead to disability, which is not only tragic on a personal level but also costly for society as a whole (Figure 7.1).

Guess what—up till now there's no high-tech scan or magical test that can tell us, "Oh yes! This is a TBI (traumatic brain injury) victim." However, hope is on the horizon! As recent findings show that blood biomarkers

 DOI: 10.1201/9781032676579-7

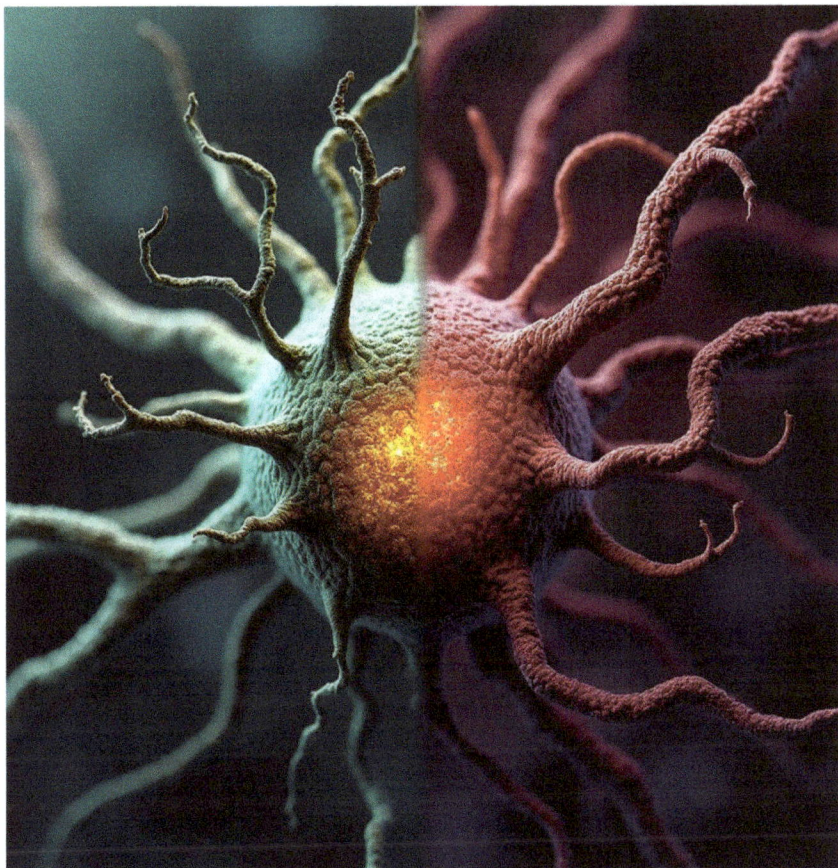

Figure 7.1 An AI-generated image of tau protein accumulation (tauopathy). Image generated using Night Cafe from the prompt "a Scientific Illustration of a Healthy Neuron vs. Affected Neuron: Side-by-Side Comparison, Healthy Neuron: Depicts Normal Tau Proteins (Green). Affected Neuron: Shows Abnormal Accumulation of Phosphorylated Tau Proteins (Red Clumps)."

could be our answer. Imagine your blood as snitching about, how much havoc the brain has undergone and even foretelling its future health status (Figure 7.2). The brain signals are like secret codes waiting for decryption—fascinating, isn't it?

Figure 7.2   An AI-generated image of a biomarker detective graphics. Image generated using Night Cafe from the prompt—Illustrate a detective character examining a blood sample under a microscope. Biomarkers symbols as clues: Representing NfL, pTau, and NSE floating in the blood sample. Background elements must include a crime scene board with connected clues, brain images, and molecular structures.

A Step-by-Step Account of Head Injuries Due to Trauma

TBI is not just a fancy word for a bad headache; TBI is a chronic disease that has been there for more than a hundred years. Back in the day it was known as dementia pugilistica, because boxers—those poor souls taking hit after hit—were often the ones who ended up with it. It is also called "punch–drunk syndrome." This is because those who continually received blows were primarily boxers like Muhammad Ali. In his earlier days, Ali was diagnosed with Parkinson's disease. He had dedicated himself to boxing, where he got repeated blows that resulted in brain injury. Today, TBI encompasses all sorts of contact sports involving athletes, soldiers, and anyone else who might have incurred multiple head traumas.

Here's another intriguing development, or should I say frightening revelation? Corsellis in 1973 examined 15 ex-boxers' brains and discovered that oligomers of certain proteins were present throughout them rather than just bruises scattered over the place. Think atrophy, holes in brain tissue, discolored areas, and even enlarged ventricles on a broader scale. At the microscopic level, it was even scarier: There was a loss of neurons; neurofibrillary tangles were developing, causing cluttered curled up strands among others as well as a destroyed blood–brain barrier system for protection. It is like an area affected by warfare without ever being cleaned.

Then what is behind all this mayhem? The leading theory is that axons—the brain's messaging cables—get torn or stretched during head trauma, triggering a cascade of destruction. These can be envisioned as thin and fragile wires delivering vital pieces of information to different areas of the human brain tissue. Such wires can be pulled, twisted, or broken when we experience some traumatic incident. When an axon breaks down, it triggers a sequence of events similar to unraveling a sweater by tugging on one string mistakenly. The damaged axons leak out chemicals that were not supposed to leak into the surrounding area, causing inflammation, which kills other nearby cells. Communication network throughout the brain is also disrupted and wide-ranging harm often results in lasting problems such as forgetfulness, mood swings, and difficulty walking.

There is one thing about TBI—it does not hit all at once. It is a slow-motion drama, or perhaps an incessant fall. Initially, these manifest as minor and temporary problems: A touch of dizziness here, some slight headache there. For instance, let us talk about Mike, who used to play soccer at college level; after suffering a number of mild traumas, he transformed into a moody person. His friends would suddenly find him laughing, and then in the next moment he was being irate over some petty thing, like losing the remote control of the TV set. It is not just a bad day; it is the early stage of TBI that will get worse if left untreated.

Stages and Symptoms of TBI

There are three stages of TBI in general (Figure 7.3):

Early Stage: Changes become evident in everyday routine, including mood swings, headache, and attempts to move smoothly during walking yet tripping. Later on cognitive functions become weak, whereas motor skills follow suit.

Late Stage: Cognitive functions deteriorate further, while motor abilities also start declining memory slips; distractedness; and such acts as going into a room and then forgetting why one went there in the first place. These and the likes pose serious challenges to the daily routine.

Advanced Stage: Late stage is now accompanied by significant brain malfunctioning, which is defined by serious cognitive, behavioral, and motor problems all at once that create conditions similar to those of advanced dementia.

TBI can also serve as an entrance to a number of other behavioral disorders, including depression, impulsivity, and suicidal tendencies. With time as cognitive impairments, escalate issues such as memory deficits or attention disorders arise prior to full-blown dementia, after which tremors and Parkinsonism ensue.

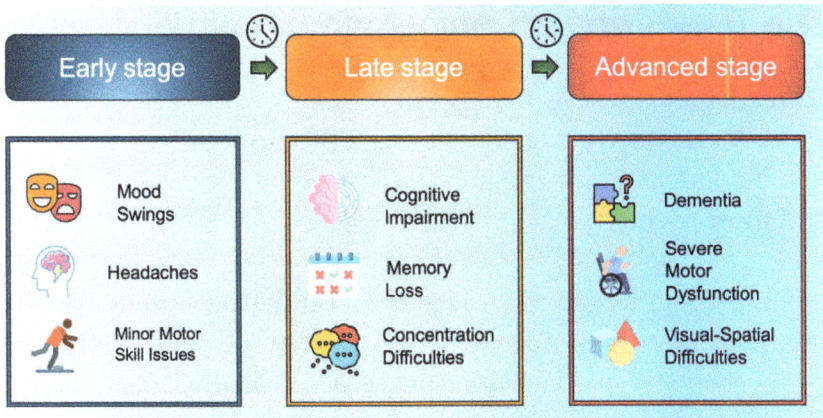

Figure 7.3   Flowchart illustrating the progression of traumatic brain injury (TBI) from the early stage to the advanced stage. Each stage highlights key symptoms and shows how TBI symptoms worsen over time.

Attribution for icons used in the illustration:

Maze icon created by Freepik: https://www.flaticon.com/free-icons/maze

Theater icon created by Good Ware: https://www.flaticon.com/free-icons/theater

Brain icon created by Kalashnyk: https://www.flaticon.com/free-icons/brain

Accident icon created by Freepik: https://www.flaticon.com/free-icons/accident

Absence icon created by Freepik: https://www.flaticon.com/free-icons/absence

Thoughts icon created by Freepik: https://www.flaticon.com/free-icons/thoughts

Missing icon created by Parzival' 1997: https://www.flaticon.com/free-icons/missing

Wheelchair icon created by Freepik: https://www.flaticon.com/free-icons/wheelchair

Geometric icon created by Freepik: https://www.flaticon.com/free-icons/geometric

The Tensegrity of Concussion: Identifying Stages

To really get down to the specifics of TBI, you must know about McKee's work on classifying TBI into four stages: the first being least dramatic and the last being the most dramatic and most frightening.

Stage One: Everything looks good on the outside; however, there are some microscopic changes. You might have occasional headaches or might occasionally experience memory lapses but nothing significant. Normally you wouldn't even realize you are sick.

Stage Two: The brain begins to show physical signs such as mild ventricular dilation (an increase in size of fluid-filled cavities within the brain) and fading in color in certain regions like substantia nigra. In this stage, tau deposits start to appear deep inside sulci of the brain, which are its microscopic folds. Symptoms start showing up more frequently: Chronic headaches that never seem to disappear and memory loss.

Stage Three: That is when things fall apart; your brain is no longer recognizable due to shrinkage (atrophy), widespread neurofibrillary tangles appear. These are made up of tau as well as amyloid plaques and are associated with Alzheimer's disease. This time you're not just forgetful—you feel mad, can never focus on one thing at a time, and frankly, not yourself. As shocking as it may be, one-third of ex-NFL players who reach this level of CTE condition commit suicide.

Stage Four: It is hopeless by now. The shrinkage has affected nearly all parts of your brain, and tau and neurofibrillary tangles are everywhere. Hypothalamus, mammillary body, and substantia nigra are among the brain's controls which if affected will show symptoms, including poor cognitive functions like executive failures to severe dementia. These will be accompanied by problems in motor coordination such as walking difficulties or even visual spatial issues.

It becomes glaringly obvious that TBI is catastrophic at this juncture. Unfortunately, early detection is not easy because there are no observable changes in the first stage. That's why scientists are searching for a blood test that can identify TBI before it becomes uncontrollably serious—a serological test, they call it (Figure 7.4).

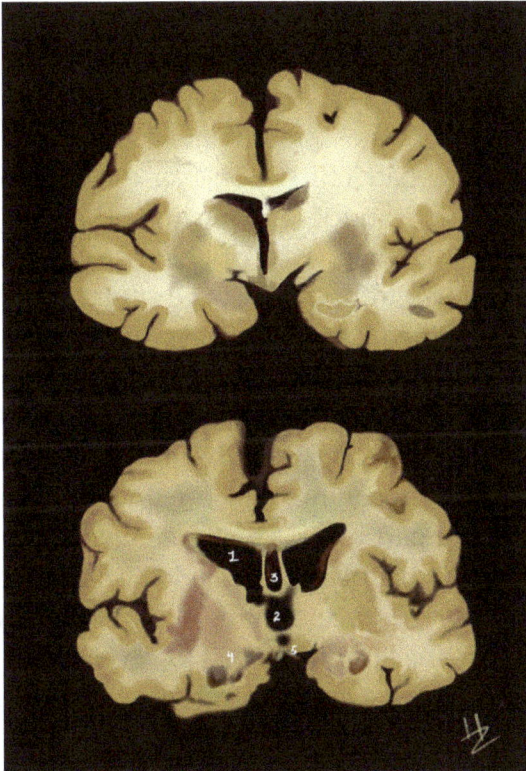

Figure 7.4 Macroscopic picture of chronic traumatic encephalopathy (CTE). 1 and 2 are dilatation of ventricles, 3 is cavum septum pellucidum, 4 is atrophy of the medial temporal lobe structures, and 5 is shrinkage of the mammillary bodies

Illustration done by Dr. Hamna Zafar, from Stern et al.'s research paper.

The Real Cost of Sports: Broken Hearts or Unbroken Heads

Anyone who watches American football, hockey, or soccer cannot help but be familiar with the term "dings," which means concussing during play. It seems like nothing much right? A slight knock on your head, shake off, and you are back into the game. However, one point must be understood—repetition of these little dings does make a big issue when summed up together. Just so you know, over 100,000–300,000 concussions occur annually among those who play American football alone. Soccer players also face the same risk; this is true for boxers, basketball players, as well as ice hockey stars. This amounts to wagering with life on one's brain (Figure 7.5).

To illustrate how bad TBI is, consider the example of Rob Kelly, a retired American footballer who had his days cut short at 28 years following years of being at the receiving end of tackles. Post-retiring from active sports however, his life became miserable in so many ways. There were wide mood swings that became very characterized, sleep became a nightmare, leading to total incapacitation. In his personal domain it became worse off because he could barely keep relationships, loss of meaning, which plunged him into deep depression. On the part of his family members too, they were no less hit by this problem; in addition to emotional stress, they had to take care of him on a full-time basis while grappling with financial difficulties. However, beyond his family members, Rob's scenario received wide criticism for revealing social costs associated with recurrent head injuries in competitive sporting activities thus accentuating why there should be better safety mechanisms as well as athletes' welfare structures. Increased awareness resulted when his family members along with other people started questioning the link between brain injuries caused by contact sports and footballers' lives.

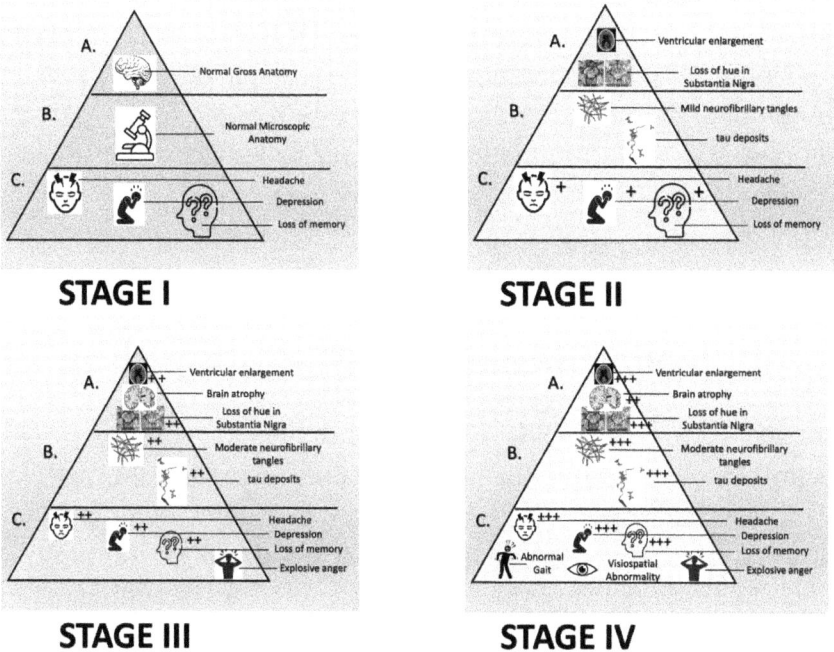

Figure 7.5   Staging of CTE. A: This row shows gross changes in the brain in each stage. B: Microscopic changes in the brain are shown in these rows in each stage. C: Clinical signs and symptoms of each stage of CTE shown in row C.

Figure taken from McKee's categorization of CTE 17.

What are some long-term effects on individuals who suffer brain damage due to trauma? According to Hillier et al., approximately 30% lose arm function, while 34% are unable to balance, leading them into wheelchairs. This means that people may not develop career-wise due to such ailments, which restrict their ability to depend on themselves or even make any effort at a career objective as a human right. First of all, there will be major health bills for families when they have members suffering from such infirmities, because if one of them is incapacitated partially or wholly, there would not only be rehabilitations but also substantial medical expenses alongside loss of wages for people who would have otherwise cared for them full time within the same household. Furthermore, these conditions result in greater pressure on national health expenditure programs like Medicare, necessitating added funds for interminable services and social service-related initiatives.

Looking for Clues in the Maze of Your Mind: Hunting Biomarkers

But let's turn now to biomarkers—those sneaky proteins swimming around in your blood or cerebrospinal fluid that could someday spell out exactly what's going wrong in your brain (Table 7.1). Picture them as distress signals for the brain, much like fire alarm notifications when there's something

Table 7.1 Biomarkers' Rubric

Criteria	Poor Indicator 1 Point	Average Indicator 2 Points	Good Indicator 3 Points
Detection in blood	Only CSF	CSF detection is more than blood	Both serum and CSF with almost equal sensitivity
Origin	Originates majorly from body organs other than brain	Originates from brain and other body organs on almost equal basis	Originates majorly from brain
Quantification sensitivity	Microlevel	Nanolevel	Picolevel
Acute and/or chronic brain injury detection	Short-term damage	Short-term and long-term damage	Long-term injury

burning somewhere around you. So, the aim is to find one biomarker—or better still, a group of them—that could diagnose TBI beyond any reasonable doubt and before it is too late.

This is where things get tricky: Blood is an ideal place where these biomarkers can be found because it is easy to extract blood. Nonetheless, blood contains too much protein, hence looking for one out of the lot is akin to searching for a pin in a haystack. Additionally, CSF contains fewer proteins, making it easier to analyze, but the only way to obtain it is via lumbar puncture—a relatively painful procedure. As for urine, it can't be usable because its biomarker concentrations are low due to other non-TBI factors that may affect them. It is just like hearing a voice message on a phone, which is full of noise, making it difficult to draw any clear conclusions.

The most promising biomarkers for TBI have been scoured by scientists who combed through decades of research before eventually ranking them using factors such as their detectability in blood, sensitivity, and origin (Figure 7.6). Three proteins emerged as the winners, and are considered the best for indicating brain injury reliably (according to the

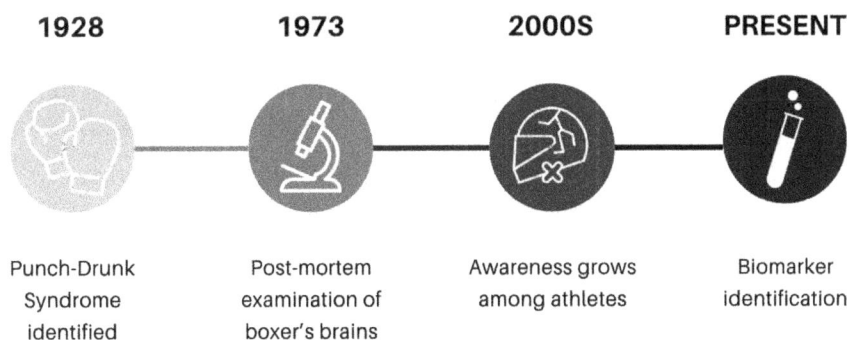

1928	1973	2000S	PRESENT
Punch-Drunk Syndrome identified	Post-mortem examination of boxer's brains	Awareness grows among athletes	Biomarker identification

Figure 7.6   Timeline illustrating key milestones in the research and understanding of traumatic brain injury (TBI), from the early identification of "punch–drunk syndrome" in 1928 to present-day advances in biomarker research (created on canva).

research on which this book is based). These proteins include neurofilament light chain (NfL), phosphorylated tau (pTau), and neuron-specific enolase (NSE).

Neurofilament light chain (NfL) is a structural protein found mainly in neurons' axons. This protein is released into the blood when the axons are damaged due to trauma, making it a very good marker for determining the extent of neuronal damage. It easily correlates with the severity of brain injury, especially in chronic conditions like CTE, where axonal injury accumulates over time. The ability to measure NfL in blood samples, which are easy to get, makes it an affordable way without a need for incisions in order to track disease progression or assess treatment efficacy.

Phosphorylated Tau (pTau) is another major player specifically in terms of chronic brain injuries such as CTE. Tau is a protein that helps stabilize microtubules in neurons but once it gets phosphorylated due to repeated head trauma, it starts clumping together into neurofibrillary tangles, which disrupt normal cell function, leading to cell death. Elevated levels of pTau within blood or cerebrospinal fluid signify ongoing neurodegeneration, providing a clear marker of the pathological processes underlying chronic traumatic brain injuries.

Neuron-specific enolase (NSE) refers to an enzyme found only inside neural cells, including nerve ends. Enzymes and other forms of proteins after they are used are usually taken up by the cells where they were synthesized, before being broken down into its constituent amino acids. When there is neuronal injury, NSE escapes into the bloodstream, indicating neuronal injury. NSE is particularly reliable when it comes to diagnosing acute brain damage but has implications in chronic conditions as well, where neurons continue to die due to multiple traumas over time. The presence of NSE in the blood renders it an important tool for estimating the gravity of brain injuries, including CTE, as well as monitoring patient outcomes longitudinally (Figure 7.7).

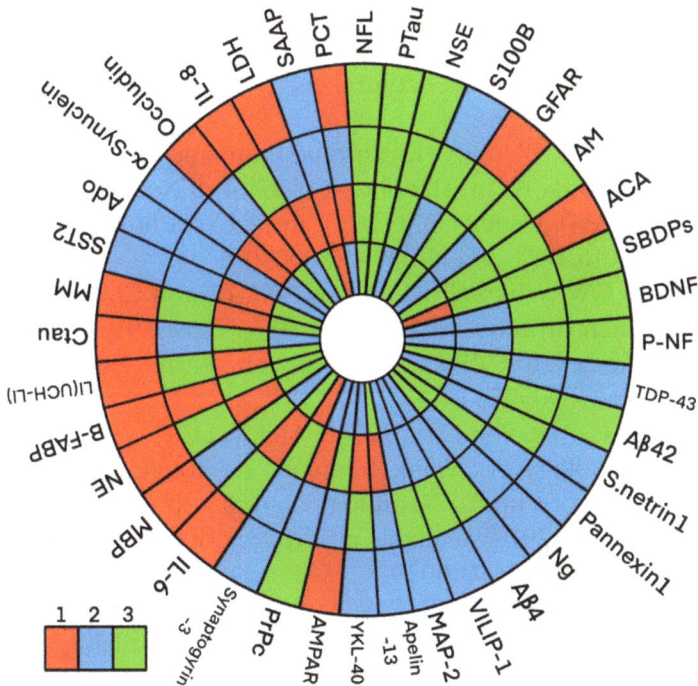

Figure 7.7 Scoring scheme of biomarkers. Red color represents score 1, blue is score 2, and green is score 3. The four rings in the circle represent the four criteria in Table 7.1. The inner most ring shows Criterion 1: Detection in blood; the second ring shows Criterion 2: Origin; the third ring shows Criterion 3: Quantification sensitivity; and the outermost ring shows Criterion 4: Brain injury detection.

Bibliography

1. Stern RA, Riley DO, Daneshvar DH, Nowinski CJ, Cantu RC, Mckee AC. Long-term consequences of repetitive brain trauma: Chronic traumatic encephalopathy. PMRJ. 2011;3(10):S460–S467. Available at http://dx.doi.org/10.1016/j.pmrj.2011.08.008.

2. Alosco ML, Mariani ML, Adler CH, Balcer LJ, Bernick C, Au R, et al. Developing methods to detect and diagnose chronic traumatic encephalopathy during life: Rationale,

design, and methodology for the DIAGNOSE CTE Research Project. Alzheimers Res. Ther. 2021;13(1):136. Available at https://alzres.biomedcentral.com/articles/10.1186/s13195-021-00872-x.

3. Zetterberg H, Blennow K. Chronic traumatic encephalopathy: Fluid biomarkers. In: Handbook of Clinical Neurology (1st ed., pp. 323–333). Elsevier B.V., 2018. Available at http://dx.doi.org/10.1016/B978-0-444-63954-7.00030-6.

4. Roberts GW, Whitwell HL, Acland PR, Bruton CJ. Dementia in a punch-drunk wife. Lancet 1990;335(8694):918–919. Available at https://linkinghub.elsevier.com/retrieve/pii/014067369090520F.

5. Mckee AC, Abdolmohammadi B, Stein TD. The neuropathology of chronic traumatic encephalopathy. In: Handbook of Clinical Neurology (1st ed., pp. 297–307). Elsevier B.V., 2018. Available at http://dx.doi.org/10.1016/B978-0-444-63954-7.00028-8.

6. Corsellis JAN, Bruton CJ, Freeman-Browne D. The aftermath of boxing. Psychol. Med. 1973;3(3):270–303. Available at http://www.embase.com/search/results?subaction=viewrecord&from=export&id=L20173271.

7. Shahim P, Gill JM, Blennow K, Zetterberg H. Fluid biomarkers for chronic traumatic encephalopathy. Semin. Neurol. 2020;40(4):411–419. Available at http://www.thieme-connect.de/DOI/DOI?10.1055/s-0040-1715095.

8. DeKosky ST, Blennow K, Ikonomovic MD, Gandy S. Acute and chronic traumatic encephalopathies: Pathogenesis and biomarkers. Nat. Rev. Neurol. 2013;9(4):192–200. Available at http://www.nature.com/articles/nrneurol.2013.36.

9. McKee AC, Cairns NJ, Dickson DW, Folkerth RD, Dirk Keene C, Litvan I, et al. The first NINDS/NIBIB consensus meeting to define neuropathological criteria for the diagnosis of chronic traumatic encephalopathy. Acta Neuropathol. 2016;131(1):75–86. Available at http://link.springer.com/10.1007/s00401-015-1515-z.

10. Gardner RC, Perry DC. Chronic traumatic encephalopathy. The Behavioral Neurology of Dementia (2nd ed., pp. 400–414). Cambridge University Press, 2022. Available at https://www.nhs.uk/conditions/chronic-traumatic-encephalopathy/ (accessed July 25, 2022).

11. Hiskens MI, Schneiders AG, Angoa-Pérez M, Vella RK, Fenning AS. Blood biomarkers for assessment of mild traumatic brain injury and chronic traumatic encephalopathy. Biomarkers 2020 2;25(3):213–227. Available at https://www.tandfonline.com/doi/full/10.1080/1354750X.2020.1735521.
12. Mawdsley C, Ferguson FR. Neurological Disease in Boxers. Lancet 1963;282(7312):795–801. Available at https://linkinghub.elsevier.com/retrieve/pii/S0140673663906408.
13. Stern RA, Daneshvar DH, Baugh CM, Seichepine DR, Montenigro PH, Riley DO, et al. Clinical presentation of chronic traumatic encephalopathy. Neurology 2013;81(13):1122–1129. Available at http://www.thieme-connect.de/DOI/DOI?10.1055/s-0040-1713624.
14. Sayed N, Culver C, Dams-O'Connor K, Hammond F, Diaz-Arrastia R. Clinical phenotype of dementia after traumatic brain injury. J Neurotrauma. 2013;30(13):1117–1122. Available at http://www.liebertpub.com/doi/10.1089/neu.2012.2638.
15. Shively S, Scher AI, Perl DP, Diaz-Arrastia R. Dementia resulting from traumatic brain injury. Arch. Neurol. 2012;69(10):1245–1251. Available at http://archneur.jamanetwork.com/article.aspx?doi=10.1001/archneurol.2011.3747.
16. Ling H, Hardy J, Zetterberg H. Neurological consequences of traumatic brain injuries in sports. Mol. Cell. Neurosci. 2015;66(PB):114–122. Available at http://dx.doi.org/10.1016/j.mcn.2015.03.012.
17. McKee AC, Daneshvar DH, Alvarez VE, Stein TD. The neuropathology of sport. Acta Neuropathol. 2014;127(1):29–51. Available at http://link.springer.com/10.1007/s00401-013-1230-6.
18. Lakhan SE, Kirchgessner A. Chronic traumatic encephalopathy: The dangers of getting "dinged." Springerplus. 2012;1(1):2. Available at http://www.springerplus.com/content/1/1/2.
19. Remembering Rob Kelly: CTE as a long-term disability. 2022, p. 6870. Available at https://www.longtermdisabilitylawyer.com/2018/04/remembering-rob-kelly-cte-as-a-long-term-disability/ (accessed July 25, 2022).
20. Hillier SL, Sharpe MH, Metzer J. Outcomes 5 years post traumatic brain injury (with further reference to neurophysical impairment and disability). Brain Inj. 1997;11(9):661–675. Available at http://www.tandfonline.com/doi/full/10.1080/026990597123214.

21. Espinoza SE, Quiben M, Hazuda HP. Distinguishing comorbidity, disability, and frailty. Curr. Geriatr. Reports 2018;7(4):201–9. Available at http://link.springer.com/10.1007/s13670-018-0254-0.

22. Tandon R. What is a biomarker? Definitions, types, and research applications explained. 2016. Available at https://bitesizebio.com/26559/biomarkers-explained/ (accessed July 25, 2022).

23. Shaw G. The use and potential of pNF-H as a general blood biomarker of axonal loss: An immediate application for CNS injury. In: Brain Neurotrauma: Molecular, Neuropsychological, and Rehabilitation Aspects. Florida: CRC Press/Taylor & Francis, 2015. Available at http://www.ncbi.nlm.nih.gov/pubmed/26269910.

24. Janigro D, Mondello S, Posti JP, Unden J. GFAP and S100B: What you always wanted to know and never dared to ask. Front. Neurol. 2022;13:835597. Available at https://www.frontiersin.org/articles/10.3389/fneur.2022.835597/full

25. Muhammad Khyani IA, Qureshi MA, Umar Farooq M, Mirza T. Molecular diagnosis of oral pre-malignant lesions & oral squamous cell carcinoma in saliva-A breakthrough in Pakistan. Int. J. Endorsing Heal. Sci. Res. 2014;2(2):108. Available at http://aeirc-edu.com/wp-content/uploads/download-full-paper101.pdf.

26. Zafari S, Backes C, Meese E, Keller A. Circulating biomarker panels in Alzheimer's disease. Gerontology 2015;61(6):497–503.

27. Arksey H, O'Malley L. Scoping studies: Towards a methodological framework. Int. J. Soc. Res. Methodol. 2005;8(1):19–32. Available at https://doi.org/10.1080/1364557032000119616.

28. Tricco AC, Lillie E, Zarin W, O'Brien KK, Colquhoun H, Levac D, et al. PRISMA extension for scoping reviews (PRISMA-ScR): Checklist and explanation. Ann. Intern. Med. 2018;169(7):467–473. Available at https://www.acpjournals.org/doi/10.7326/M18-0850.

29. Brookhart SM. Appropriate criteria: Key to effective rubrics. Front. Educ. 2018;3. Available at http://journal.frontiersin.org/article/10.3389/feduc.2018.00022/full.

30. Garban Z. Biomarkers: Theoretical aspects and applicative peculiarities note II. Nutritional biomarkers. J. Agroaliment Process Technol. 2006;XII(2):349–356.

31. Robinson S, Pool R, Giffin R. Qualifying biomarkers. Emerging Safety Science: Workshop Summary. 2008.

32. Anderson KJ, Scheff SW, Miller KM, Roberts KN, Gilmer LK, Yang C, et al. The phosphorylated axonal form of the neurofilament subunit NF-h (pNF-h) as a blood biomarker of traumatic brain injury. J. Neurotrauma. 2008;25(9): 1079–1085.

33. Ghonemi MO, Rabah AA, Saber HM, Radwan W. Role of phosphorylated neurofilament h as a diagnostic and prognostic marker in traumatic brain injury. Egypt J. Crit Care Med. 2013; 1(3):139–144. Available at http://dx.doi.org/10.1016/j.ejccm.2013. 03.002.

34. Shahim P, Politis A, van der Merwe A, Moore B, Ekanayake V, Lippa SM, et al. Time course and diagnostic utility of NfL, tau, GFAP, and UCH-L1 in subacute and chronic TBI. Neurology 2020;95(6):e623–e636.

35. Hampel H, Teipel SJ. Total and phosphorylated tau proteins: Evaluation as core biomarker candidates in frontotemporal dementia. Dement. Geriatr. Cogn. Disord. 2004;17(4):350–4. Available at http://dx.doi.org/10.1038/s41467-021-23620-z.

36. Ding X, Zhang S, Jiang L, Wang L, Li T, Lei P. Ultrasensitive assays for detection of plasma tau and phosphorylated tau 181 in Alzheimer's disease: A systematic review and meta-analysis. Transl. Neurodegener. 2021;10(1):1–14.

37. Thelin EP, Jeppsson E, Frostell A, Svensson M, Mondello S, Bellander BM, et al. Utility of neuron-specific enolase in traumatic brain injury; relations to S100B levels, outcome, and extracranial injury severity. Crit Care. 2016;20(1):1–15. Available at http://dx.doi.org/10.1186/s13054-016-1450-y.

38. Kacira T, Kemerdere R, Atukeren P, Hanimoglu H, Sanus GZ, Kucur M, et al. Detection of caspase-3, neuron specific enolase, and high-sensitivity C-reactive protein levels in both cerebrospinal fluid and serum of patients after aneurysmal subarachnoid hemorrhage. Neurosurgery 2007;60(4):674–9.

39. Agoston DV, Shutes-David A, Peskind ER. Biofluid biomarkers of traumatic brain injury. Brain Inj. 2017;31(9):1195–203. Available at https://doi.org/10.1080/0269 9052.2017.1357836.

40. Bersani I, Pluchinotta F, Dotta A, Savarese I, Campi F, Auriti C, et al. Early predictors of perinatal brain damage: The role of neurobiomarkers. Clin. Chem. Lab Med. 2020;58(4):471–486.

41. Thelin EP, Nelson DW, Bellander BM. A review of the clinical utility of serum S100B protein levels in the assessment of traumatic brain injury. Acta Neurochir (Wien). 2017;159(2):209–225.

42. Undén J, Bellner J, Eneroth M, Alling C, Ingebrigtsen T, Romner B. Raised serum S100B levels after acute bone fractures without cerebral injury. J. Trauma. Inj. Infect. Crit. Care 2005;58(1):59–61.

43. DeDominicis KE, Hwang H, Cartagena CM, Shear DA, Boutté AM. Cerebrospinal fluid biomarkers are associated with glial fibrillary acidic protein and αII-spectrin breakdown products in brain tissues following penetrating ballistic-like brain injury in rats. Front. Neurol. 2018;9:1–14.

44. Jones A, Jarvis P. Review of the potential use of blood neuro-biomarkers in the diagnosis of mild traumatic brain injury. Clin. Exp. Emerg. Med. 2017;4(3):121–127.

45. Katsanos AH, Makris K, Stefani D, Koniari K, Gialouri E, Lelekis M, et al. Plasma glial fibrillary acidic protein in the differential diagnosis of intracerebral hemorrhage. Stroke 2017;48(9):2586–2588.

46. Robertson CL, Marion DW, Kochanek PM, Minamino N, Ruppel RA, Kangawa K, et al. Increased adrenomedullin in cerebrospinal fluid after traumatic brain injury in infants and children. J. Neurotrauma. 2001;18(9):861–868.

47. Shan J, Stachniak T, Jhamandas JH, Krukoff TL. Autonomic and neuroendocrine actions of adrenomedullin in the brain: Mechanisms for homeostasis. Regul Pept. 2003;112(1–3):33–40.

48. Xue LX, Liu HY, Cui Y, Dong Y, Wang JQ, Ji QY, et al. Neuroprotective effects of activin A on endoplasmic reticulum stress-mediated apoptotic and autophagic PC12 cell death. Neural. Regen. Res. 2017;12(5):779–786.

49. Pike BR, Flint J, Dutta S, Johnson E, Wang KKW, Hayes RL. Accumulation of non-erythroid αII-spectrin and calpain-cleaved αII-spectrin breakdown products in cerebrospinal fluid after traumatic brain injury in rats. J. Neurochem. 2001;78(6):1297–306.

50. Failla MD, Conley YP, Wagner AK. Brain-derived neurotrophic factor (BDNF) in traumatic brain injury–related mortality. Neurorehabil. Neural Repair 2016;30(1):83–93.

51. Wu GWY, Wolkowitz OM, Reus VI, Kang JI, Elnar M, Sarwal R, et al. Serum brain-derived neurotrophic factor remains elevated after long term follow-up of combat veterans with chronic post-traumatic stress disorder. Psychoneuroendocrinology 2021;134:105360.

52. Feneberg E, Steinacker P, Lehnert S, Schneider A, Walther P, Thal DR, et al. Limited role of free TDP-43 as a diagnostic tool in neurodegenerative diseases. Amyotroph. Lateral Scler. Front. Degener. 2014;15(5–6):351–356. Available at http://www.tandfonline.com/doi/full/10.3109/21678421.2014.905606.

53. Johnson VE, Stewart W, Trojanowski JQ, Smith DH. Acute and chronically increased immunoreactivity to phosphorylation-independent but not pathological TDP-43 after a single traumatic brain injury in humans. Acta Neuropathol. 2011;122(6):715–726. Available at http://link.springer.com/10.1007/s00401-011-0909-9.

54. Suk TR, Rousseaux MWC. The role of TDP-43 mislocalization in amyotrophic lateral sclerosis. Mol. Neurodegener. 2020;15(1):45. Available at https://molecularneurodegeneration.biomedcentral.com/articles/10.1186/s13024-020-00397-1.

55. West T, Kirmess KM, Meyer MR, Holubasch MS, Knapik SS, Hu Y, et al. A blood-based diagnostic test incorporating plasma Aβ42/40 ratio, ApoE proteotype, and age accurately identifies brain amyloid status: Findings from a multi cohort validity analysis. Mol. Neurodegener. 2021;16(1):30. Available at https://molecularneurodegeneration.biomedcentral.com/articles/10.1186/s13024-021-00451-6.

56. Chen T-B, Lee Y-J, Lin S-Y, Chen J-P, Hu C-J, Wang P-N, et al. Plasma Aβ42 and total tau predict cognitive decline in amnestic mild cognitive impairment. Sci. Rep. 2019;9(1):13984. Available at http://dx.doi.org/10.1038/s41598-019-50315-9.

57. Xie Y, Guo Z, Chen F, Xiao C, Xu J, Bo D. Serum netrin-1 as a potential biomarker for functional outcome of traumatic brain injury. Clin. Chim. Acta 2021;518:22–27. Available at https://linkinghub.elsevier.com/retrieve/pii/S000989812100084X.

58. Guo D, Zhu Z, Zhong C, Peng H, Wang A, Xu T, et al. Increased serum netrin-1 is associated with improved prognosis of ischemic stroke. Stroke 2019;50(4):845–852. Available at https://www.ahajournals.org/doi/10.1161/STROKEAHA.118.024631.

59. Yim J, Kim G, Lee B-W, Kang ES, Cha B-S, Kim J-H, et al. Relationship between circulating netrin-1 concentration, impaired fasting glucose, and newly diagnosed type 2 diabetes. Front. Endocrinol. (Lausanne). 2018;9:1–8. Available at https://www.frontiersin.org/article/10.3389/fendo.2018.00691/full.

60. Koval M, Cwiek A, Carr T, Good ME, Lohman AW, Isakson BE. Pannexin 1 as a driver of inflammation and ischemia–reperfusion injury. Purinergic Signal. 2021;17(4):521–531. Available at https://doi.org/10.1007/s11302-021-09804-8.

61. Human PANX1(Pannexin-1) ELISA Kit. 2022. Available at https://www.fn-test.com/product/eh1283/ (accessed July 22, 2022).

62. Zhang Y, Laumet G, Chen S-R, Hittelman WN, Pan H-L. Pannexin-1 up-regulation in the dorsal root ganglion contributes to neuropathic pain development. J. Biol. Chem. 2015;290(23):14647–14655. Available at https://linkinghub.elsevier.com/retrieve/pii/S0021925819795246.

63. DeLalio LJ, Masati E, Mendu S, Ruddiman CA, Yang Y, Johnstone SR, et al. Pannexin 1 channels in renin-expressing cells influence renin secretion and blood pressure homeostasis. Kidney Int. 2020;98(3):630–644. Available at https://linkinghub.elsevier.com/retrieve/pii/S0085253820305433.

64. Yang J, Korley FK, Dai M, Everett AD. Serum neurogranin measurement as a biomarker of acute traumatic brain injury. Clin. Biochem. 2015;48(13–14):843–848. Available at https://linkinghub.elsevier.com/retrieve/pii/S0009912015002064.

65. Xiang Y, Xin J, Le W, Yang Y. Neurogranin: A potential biomarker of neurological and mental diseases. Front. Aging Neurosci. 2020;12:584743. Available at https://www.frontiersin.org/article/10.3389/fnagi.2020.584743/full.

66. Marklund N, Farrokhnia N, Hånell A, Vanmechelen E, Enblad P, Zetterberg H, et al. Monitoring of β-amyloid dynamics after human traumatic brain injury. J. Neurotrauma. 2014;31(1):42–55. Available at http://www.liebertpub.com/doi/10.1089/neu.2013.2964.

67. Emmerling MR, Morganti-Kossmann MC, Kossmann T, Stahel PF, Watson MD, Evans LM, et al. Traumatic brain injury elevates the Alzheimer's amyloid peptide Abeta42 in human CSF: A possible role for nerve cell injury. Ann N. Y. Acad. Sci.

2000;903(1):118–122. Available at https://onlinelibrary.wiley.com/doi/10.1111/j.1749-6632.2000.tb06357.x.

68. Vanderstichele H, Kerschaver E, Van, Hesse C, Davidsson P, Buyse M-A, Andreasen N, et al. Standardization of measurement of β-amyloid (1–42) in cerebrospinal fluid and plasma. Amyloid 2000;7(4):245–258. Available at http://www.tandfonline.com/doi/full/10.3109/13506120009146438.

69. Roher AE, Esh CL, Kokjohn TA, Castaño EM, Van Vickle GD, Kalback WM, et al. Amyloid beta peptides in human plasma and tissues and their significance for Alzheimer's disease. Alzheimer's Dement. 2009;5(1):18–29. Available at https://onlinelibrary.wiley.com/doi/10.1016/j.jalz.2008.10.004.

70. Hasegawa S, Matsushige T, Inoue H, Takahara M, Kajimoto M, Momonaka H, et al. Serum and cerebrospinal fluid levels of visinin-like protein-1 in acute encephalopathy with biphasic seizures and late reduced diffusion. Brain Dev. 2014;36(7):608–612. Available at http://dx.doi.org/10.1016/j.braindev.2013.08.010.

71. Tarawneh R, D'Angelo G, MacY E, Xiong C, Carter D, Cairns NJ, et al. Visinin-like protein-1: Diagnostic and prognostic biomarker in Alzheimer's disease. Ann. Neurol. 2011;70(2):274–285.

72. Tan Z, Jiang J, Tian F, Peng J, Yang Z, Li S, et al. Serum visinin-like protein 1 is a better biomarker than neuron-specific enolase for seizure-induced neuronal injury: A prospective and observational study. Front. Neurol. 2020;11:1–7.

73. Braunewell K-H, Szanto AJK. Visinin-like proteins (VSNLs): Interaction partners and emerging functions in signal transduction of a subfamily of neuronal Ca^{2}+-sensor proteins. Cell Tissue Res. 2009;335(2):301–316. Available at http://link.springer.com/10.1007/s00441-008-0716-3.

74. Fu J, Jin F, Zhang J, Fong K, Bassi DE, Lopez De Cicco R, et al. VILIP-1 expression in vivo results in decreased mouse skin keratinocyte proliferation and tumor development. PLoS One. 2010;5(4):e10196. Available at https://dx.plos.org/10.1371/journal.pone.0010196.

75. Taft WC, Yang K, Dixon CE, Clifton GL, Hayes RL. Hypothermia attenuates the loss of hippocampal microtubule-associated protein 2 (MAP2) following traumatic brain injury. J. Cereb. Blood Flow Metab. 1993;13(5):796–802. Available at http://journals.sagepub.com/doi/10.1038/jcbfm.1993.101.

76. Mondello S, Hayes RL Biomarkers. Handb. Clin. Neurol. 2015;127:245–265. Available at https://linkinghub.elsevier.com/retrieve/pii/B9780444528926000167.

77. Mages B, Fuhs T, Aleithe S, Blietz A, Hobusch C, Härtig W, et al. The cytoskeletal elements MAP2 and NF-L show substantial alterations in different stroke models while elevated serum levels highlight especially MAP2 as a sensitive biomarker in stroke patients. Mol. Neurobiol. 2021;58(8):4051–4069. Available at https://link.springer.com/10.1007/s12035-021-02372-3.

78. Park D, Joo SS, Lee HJ, Choi K, Kim SU, Kim Y. Microtubule-associated protein 2, an early blood marker of ischemic brain injury. J. Neurosci. Res. 2012;90(2):461–467. Available at https://onlinelibrary.wiley.com/doi/10.1002/jnr.22769.

79. Luo H, Xiang Y, Qu X, Liu H, Liu C, Li G, et al. Apelin-13 suppresses neuroinflammation against cognitive deficit in a streptozotocin-induced rat model of Alzheimer's disease through activation of BDNF-TrkB signaling pathway. Front. Pharmacol. 2019;10:1–14. Available at https://www.frontiersin.org/article/10.3389/fphar.2019.00395/full.

80. Xu W, Li T, Gao L, Zheng J, Yan J, Zhang J, et al. Apelin-13/APJ system attenuates early brain injury via suppression of endoplasmic reticulum stress-associated TXNIP/NLRP3 inflammasome activation and oxidative stress in a AMPK-dependent manner after subarachnoid hemorrhage in rats. J. Neuroinflammation 2019;16(1):247. Available at https://jneuroinflammation.biomedcentral.com/articles/10.1186/s12974-019-1620-3.

81. Yildiz Z, Eren N, Orcun A, Münevver Gokyigit F, Turgay F, Gündogdu Celebi L. Serum apelin-13 levels and total oxidant/antioxidant status of patients with Alzheimer's disease. AGING Med. 2021;4(3):201–205. Available at https://onlinelibrary.wiley.com/doi/10.1002/agm2.12173.

82. Alpua M, Turkel Y, Dag E, Kisa U. Apelin-13: A promising biomarker for multiple sclerosis? Ann. Indian Acad. Neurol. 2018;21(2):126. Available at http://www.annalsofian.org/text.asp?2018/21/2/126/237155.

83. Zhao T, Su Z, Li Y, Zhang X, You Q. Chitinase-3 like-protein-1 function and its role in diseases. Signal Transduct. Target Ther. 2020;5(1):201. Available at http://dx.doi.org/10.1038/s41392-020-00303-7.

84. Human YKL-40 ELISA Kit (ab255719), Vol. 1130. 2008. pp. 3–5.
85. Bonneh-Barkay D, Zagadailov P, Zou H, Niyonkuru C, Figley M, Starkey A, et al. YKL-40 expression in traumatic brain injury: An initial analysis. J. Neurotrauma. 2010;27(7):1215–23. Available at http://www.liebertpub.com/doi/10.1089/neu.2010.1310.
86. Baldacci F, Lista S, Cavedo E, Bonuccelli U, Hampel H. Diagnostic function of the neuroinflammatory biomarker YKL-40 in Alzheimer's disease and other neurodegenerative diseases. Expert Rev Proteomics. 2017;14(4):285–299. Available at https://www.tandfonline.com/doi/full/10.1080/14789450.2017.1304217.
87. Yokoi N, Fukata M, Fukata Y. Synaptic plasticity regulated by protein–protein interactions and posttranslational modifications. Int. Rev. Cell Mol. Biol. 2012;297:1–43. Available at http://dx.doi.org/10.1016/B978-0-12-394308-8.00001-7.
88. AMPA Receptor Antibody. 2022. Available at http://www.clinlabnavigator.com/ampa-receptor-antibody.html (accessed July 13, 2022).
89. Dambinova SA, Shikuev AV, Weissman JD, Mullins JD. AMPAR peptide values in blood of nonathletes and club sport athletes with concussions. Mil. Med. 2013;178(3):285–290. Available at https://academic.oup.com/milmed/article/178/3/285-290/4210847.
90. Wulf M-A, Senatore A, Aguzzi A. The biological function of the cellular prion protein: An update. BMC Biol. 2017;15(1):34. Available at http://bmcbiol.biomedcentral.com/articles/10.1186/s12915-017-0375-5.
91. Castle AR, Gill AC. Physiological functions of the cellular prion protein. Front. Mol. Biosci. 2017;4:1–25. Available at http://journal.frontiersin.org/article/10.3389/fmolb.2017.00019/full.
92. Minikel EV, Kuhn E, Cocco AR, Vallabh SM, Hartigan CR, Reidenbach AG, et al. Domain-specific quantification of prion protein in cerebrospinal fluid by targeted mass spectrometry. Mol. Cell Proteomics. 2019;18(12):2388–2400. Available at http://dx.doi.org/10.1074/mcp.RA119.001702.
93. Poturnayova A, Snejdarkova M, Babelova L, Korri-Youssoufi H, Hianik T. Comparative analysis of cellular prion detection by mass-sensitive immunosensors. Electroanalysis 2014;26(6):1312–1319. Available at https://onlinelibrary.wiley.com/doi/10.1002/elan.201400049.

94. Goetzl EJ, Ledreux A, Granholm A-C, Elahi FM, Goetzl L, Hiramoto J, et al. Neuron-derived exosome proteins may contribute to progression from repetitive mild traumatic brain injuries to chronic traumatic encephalopathy. Front. Neurosci. 2019;13:1–8. Available at https://www.frontiersin.org/article/10.3389/fnins.2019.00452/full.

95. Belizaire R, Komanduri C, Wooten K, Chen M, Thaller C, Janz R. Characterization of synaptogyrin 3 as a new synaptic vesicle protein. J. Comp. Neurol. 2004;470(3):266–281. Available at https://onlinelibrary.wiley.com/doi/10.1002/cne.20008.

96. Synaptogyrin 3. The Human Protein Atlas, 2021, pp. 30–42.

97. Edwards KA, Gill JM, Pattinson CL, Lai C, Brière M, Rogers NJ, et al. Interleukin-6 is associated with acute concussion in military combat personnel. BMC Neurol. 2020;20(1):209. Available at https://bmcneurol.biomedcentral.com/articles/10.1186/s12883-020-01760-x.

98. Zhang J, Hao Y, Ou W, Ming F, Liang G, Qian Y, et al. Serum interleukin-6 is an indicator for severity in 901 patients with SARS-CoV-2 infection: A cohort study. J. Transl. Med. 2020;18(1):406. Available at https://doi.org/10.1186/s12967-020-02571-x.

99. Interleukin-6. The Human Protein Atlas, 2022, pp. 27–28. Available at https://www.proteinatlas.org/ENSG00000136244-IL6 (accessed July 14, 2022).

100. Erta M, Quintana A, Hidalgo J. Interleukin-6, a major cytokine in the central nervous system. Int. J. Biol. Sci. 2012;8(9):1254–1266. Available at http://www.ijbs.com/v08p1254.htm.

101. Glushakova OY, Glushakov AV, Mannix R, Miller ER, Valadka AB, Hayes RL. The use of blood-based biomarkers to improve the design of clinical trials of traumatic brain injury. In: Handbook of Neuroemergency Clinical Trials (pp. 139–166). Elsevier, 2018. Available at https://linkinghub.elsevier.com/retrieve/pii/B9780128040645000084.

102. Singh A, Singh K, Sahu A, Prasad RS, Pandey N, Dhar S. Serum concentration of myelin basic protein as a prognostic marker in mild-to-moderate head injury patients: A prospective study in a tertiary care center. Indian J. Neurosurg. 2021;11. Available at http://www.thieme-connect.de/DOI/DOI?10.1055/s-0040-1716936.

103. Dunn-Meynell AA, Hassanain M, Levin BE. Norepinephrine and traumatic brain injury: A possible role in post-traumatic edema. Brain Res. 1998;800(2):245–252. Available at https://linkinghub.elsevier.com/retrieve/pii/S0006899398005241.

104. Roy A, Pickar D, Jong JD, Karoum F, Linnoila M. Norepinephrine and its metabolites in cerebrospinal fluid, plasma, and urine. Arch. Gen. Psychiatry 1988;45(9):849. Available at http://archpsyc.jamanetwork.com/article.aspx?doi=10.1001/archpsyc.1988.01800330081010.

105. Norepinephrine. 2022. Available at https://www.acs.org/content/acs/en/molecule-of-the-week/archive/n/norepinephrine.html (accessed July 23, 2022).

106. Denfeld QE, Habecker BA, Woodward WR. Measurement of plasma norepinephrine and 3,4-dihydroxyphenylglycol: Method development for a translational research study. BMC Res. Notes 2018;11(1):248. Available at https://doi.org/10.1186/s13104-018-3352-3.

107. Pelsers MMAL, Hanhoff T, Van der Voort D, Arts B, Peters M, Ponds R, et al. Brain- and heart-type fatty acid-binding proteins in the brain: Tissue distribution and clinical utility. Clin. Chem. 2004;50(9):1568–1575. Available at https://academic.oup.com/clinchem/article/50/9/1568/5640128

108. Pelsers MMAL, Glatz JFC. Detection of brain injury by fatty acid-binding proteins. Clin Chem Lab Med. 2005 Jan 1;43(8):802–9. Available at https://www.degruyter.com/document/doi/10.1515/CCLM.2005.135/html.

109. Anderson TN, Hwang J, Munar M, Papa L, Hinson HE, Vaughan A, et al. Blood-based biomarkers for prediction of intracranial hemorrhage and outcome in patients with moderate or severe traumatic brain injury. J. Trauma Acute Care Surg. 2020;89(1):80–86.

110. Ramezani F, Bahrami-Amiri A, Babahajian A, Shahsavari Nia K, Yousefifard M. Ubiquitin c-terminal hydrolase-L1 (UCH-L1) in prediction of computed tomography findings in traumatic brain injury: A meta-analysis. Emerg (Tehran). 2018;6(1):e62. Available at http://www.ncbi.nlm.nih.gov/pubmed/30788389%0Ahttp://www.pubmedcentral.nih.gov/articlerender.fcgi?artid=PMC6368936.

111. Gabbita SP, Scheff SW, Menard RM, Roberts K, Fugaccia I, Zemlan FP. Cleaved-tau: A biomarker of neuronal damage

after traumatic brain injury. J. Neurotrauma. 2005;22(1):83–94. Available at http://www.liebertpub.com/doi/10.1089/neu.2005.22.83.

112. Forouzan A, Motamed H, Delirrooyfard A, Zallaghi S. Serum cleaved tau protein and clinical outcome in patients with minor head trauma. Open Access Emerg Med. 2020;12:7–12. Available at https://www.dovepress.com/serum-cleaved-tau-protein-and-clinical-outcome-in-patients-with-minor–peer-reviewed-article-OAEM.

113. Cengiz P, Zemlan F, Eickhoff JC, Ellenbogen R, Zimmerman JJ. Increased cerebrospinal fluid cleaved tau protein (C-tau) levels suggest axonal damage in pediatric patients with brain tumors. Child's Nerv. Syst. 2015;31(8):1313–1319. Available at http://link.springer.com/10.1007/s00381-015-2705-7.

114. Lionnet A, Wade MA, Corbillé A-G, Prigent A, Paillusson S, Tasselli M, et al. Characterisation of tau in the human and rodent enteric nervous system under physiological conditions and in tauopathy. Acta Neuropathol Commun. 2018;6(1):65. Available at https://actaneurocomms.biomedcentral.com/articles/10.1186/s40478-018-0568-3.

115. Hayashi T, Kaneko Y, Yu S, Bae E, Stahl CE, Kawase T, et al. Quantitative analyses of matrix metalloproteinase activity after traumatic brain injury in adult rats. Brain Res. 2009;1280:172–177. Available at http://dx.doi.org/10.1016/j.brainres.2009.05.040.

116. Guilfoyle MR, Carpenter KLH, Helmy A, Pickard JD, Menon DK, Hutchinson PJA. Matrix metalloproteinase expression in contusional traumatic brain injury: A paired microdialysis study. J. Neurotrauma. 2015;32(20):1553–1559. Available at http://www.liebertpub.com/doi/10.1089/neu.2014.3764.

117. Brennan M, Bhatti H, Nerusu KC, Bhagavathula N, Kang S, Fisher GJ, et al. Matrix metalloproteinase-1 is the major collagenolytic enzyme responsible for collagen damage in UV-irradiated human skin. Photochem. Photobiol. 2003;78(1):43. Available at http://doi.wiley.com/10.1562/0031-8655(2003)078%3C0043:MMITMC%3E2.0.CO;2.

118. Lindner D, Zietsch C, Becher PM, Schulze K, Schultheiss H-P, Tschöpe C, et al. Differential expression of matrix metalloproteases in human fibroblasts with different origins. Biochem. Res. Int. 2012;2012:1–10. Available at https://www.hindawi.com/journals/bri/2012/875742/.

119. Weir RAP, Miller AM, Murphy GEJ, Clements S, Steedman T, Connell JMC, et al. Serum soluble ST2. J. Am. Coll. Cardiol. 2010;55(3):243–250. Available at https://linkinghub.elsevier.com/retrieve/pii/S0735109709035281.

120. Du Q, Weng J-F, Luo L-F, Cen M, Yu W-H, Zheng Y-K, et al. Serum ST2 as a potential prognostic biomarker for traumatic brain injury. Clin. Chim. Acta. 2018;487:145–152. Available at https://linkinghub.elsevier.com/retrieve/pii/S0009898118305217.

121. Lissak IA, Zafar SF, Westover MB, Schleicher RL, Kim JA, Leslie-Mazwi T, et al. Soluble ST2 is associated with new epileptiform abnormalities following nontraumatic subarachnoid hemorrhage. Stroke 2020;51(4):1128–1134. Available at https://www.ahajournals.org/doi/10.1161/STROKEAHA.119.028515.

122. Borea PA, Gessi S, Merighi S, Vincenzi F, Varani K. Pathological overproduction: The bad side of adenosine. Br. J. Pharmacol. 2017;174(13):1945–1960. Available at https://onlinelibrary.wiley.com/doi/10.1111/bph.13763.

123. Peters B. Adenosine and sleep. 2022. Available at https://www.verywellhealth.com/adenosine-and-sleep-3015337 (accessed July 24, 2022).

124. Löfgren L, Pehrsson S, Hägglund G, Tjellström H, Nylander S. Accurate measurement of endogenous adenosine in human blood. PLoS One 2018;13(10):e0205707. Available at https://dx.plos.org/10.1371/journal.pone.0205707.

125. Cho B-H, Kim BC, Yoon G-J, Choi S-M, Chang J, Lee S-H, et al. Adenosine deaminase activity in cerebrospinal fluid and serum for the diagnosis of tuberculous meningitis. Clin. Neurol. Neurosurg. 2013;115(9):1831–1836. Available at http://dx.doi.org/10.1016/j.clineuro.2013.05.017.

126. Karmouty-Quintana H, Xia Y, Blackburn MR. Adenosine signaling during acute and chronic disease states. J. Mol. Med. 2013;91(2):173–181. Available at http://link.springer.com/10.1007/s00109-013-0997-1.

127. Tinsley RB, Kotschet K, Modesto D, Ng H, Wang Y, Nagley P, et al. Sensitive and specific detection of α-synuclein in human plasma. J. Neurosci. Res. 2010;88(12):2693–2700. Available at https://onlinelibrary.wiley.com/doi/10.1002/jnr.22417.

128. Su E, Bell MJ, Wisniewski SR, Adelson PD, Janesko-Feldman KL, Salonia R, et al. α-Synuclein levels are elevated in cerebrospinal fluid following traumatic brain injury in infants

and children: The effect of therapeutic hypothermia. Dev. Neurosci. 2010;32(5–6):385–395. Available at https://www. karger.com/Article/FullText/321342.

129. Shi M, Liu C, Cook TJ, Bullock KM, Zhao Y, Ginghina C, et al. Plasma exosomal α-synuclein is likely CNS-derived and increased in Parkinson's disease. Acta Neuropathol. 2014;128(5):639–650. Available at http://link.springer. com/10.1007/s00401-014-1314-y.

130. Chang C-W, Yang S-Y, Yang C-C, Chang C-W, Wu Y-R. Plasma and serum alpha-synuclein as a biomarker of diagnosis in patients with Parkinson's disease. Front. Neurol. 2020;10:1–7. Available at https://www.frontiersin.org/article/10.3389/ fneur.2019.01388/full.

131. Pan R, Yu K, Weatherwax T, Zheng H, Liu W, Liu KJ. Blood occludin level as a potential biomarker for early blood brain barrier damage following ischemic stroke. Sci. Rep. 2017;7(1):40331. Available at http://dx.doi.org/10.1038/ srep40331.

132. Beeman NE, Baumgartner HK, Webb PG, Schaack JB, Neville MC. Disruption of occludin function in polarized epithelial cells activates the extrinsic pathway of apoptosis leading to cell extrusion without loss of transepithelial resistance. BMC Cell Biol. 2009;10(1):85. Available at https://bmcmolcellbiol. biomedcentral.com/articles/10.1186/1471-2121-10-85.

133. Cummins PM. Occludin: One protein, many forms. Mol. Cell Biol. 2012;32(2):242–250. Available at https://journals.asm.org/ doi/10.1128/MCB.06029-11.

134. Occludin ELISA Kit. 2022. Available at https://www.antibodies-online.com/kit/6958292/Occludin+OCLN+ELISA+Kit/ (accessed July 21, 2022).

135. Mussack T, Biberthaler P, Kanz K-G, Wiedemann E, Gippner-Steppert C, Mutschler W, et al. Serum s-100B and interleukin-8 as predictive markers for comparative neurologic outcome analysis of patients after cardiac arrest and severe traumatic brain injury. Crit. Care Med. 2002;30(12):2669–2674. Available at http://journals.lww. com/00003246-200212000-00010.

136. Remick DG. Interleukin-8. Crit Care Med. 2005;33(Suppl.):S466–S477. Available at http://journals.lww. com/00003246-200512001-00021.

137. Halstensen A, Ceska M, Brandtzaeg P, Redl H, Naess A, Waage A. Interleukin-8 in serum and cerebrospinal fluid from patients with meningococcal disease. J. Infect. Dis. 1993;167(2):471–475. Available at https://academic.oup.com/jid/article-lookup/doi/10.1093/infdis/167.2.471.

138. Alkhatib AJ. Lactate dehydrogenase: Physiological roles and clinical implications. Am. J. Biomed. Sci. Res. 2019;3(5):415–416. Available at https://biomedgrid.com/fulltext/volume3/lactate-dehydrogenase-physiological-roles-and-clinical-implications.000705.php.

139. SS HSinha MK, Prasad M, Agrawal R, Singh A, Kumar U, et al. Clinical status of lactate dehydrogenase activity in cerebrospinal fluid with age and sex distribution in different types of meningitis. MOJ Immunol. 2016;4(5):10–12. Available at https://medcraveonline.com/MOJI/clinical-status-of-lactate-dehydrogenase-activity-in-cerebrospinal-fluid-with-age-and-sex-distribution-in-different-types-of-meningitis.html.

140. Lactate Dehydrogenase (LDH). 2022. Available at https://labpedia.net/lactate-dehydrogenase-ldh-isoenzymes-of-ldh/ (accessed July 24, 2022).

141. LDH elisa kit:: Human lactate dehydrogenase ELISA Kit, p. 858. MyBiosource. Available at https://www.mybiosource.com/ldh-human-elisa-kits/lactate-dehydrogenase/9535 (accessed July 24, 2022).

142. Amyloid A protein. SpringerReference, 2022. Available at http://www.springerreference.com/index/doi/10.1007/SpringerReference_37425 (accessed July 15, 2022).

143. Yamada T, Nomata Y, Sugita O, Okada M. A rapid method for measuring serum amyloid a protein by latex agglutination nephelometric immunoassay. Ann. Clin. Biochem. 1993;30(1):72–76. Available at http://journals.sagepub.com/doi/10.1177/000456329303000112.

144. Wicker E, Benton L, George K, Furlow W, Villapol S. Serum amyloid a protein as a potential biomarker for severity and acute outcome in traumatic brain injury. Biomed. Res. Int. 2019;2019:1–9. Available at https://www.hindawi.com/journals/bmri/2019/5967816/.

145. Urieli-Shoval S, Cohen P, Eisenberg S, Matzner Y. Widespread expression of serum amyloid A in histologically normal human tissues: Predominant localization to the

epithelium. J. Histochem. Cytochem. 1998;46(12):1377–1384. Available at http://journals.sagepub.com/doi/10.1177/002215549804601206.

146. Alons IME, Verheul RJ, Kuipers I, Jellema K, Wermer MJH, Algra A, et al. Procalcitonin in cerebrospinal fluid in meningitis: A prospective diagnostic study. Brain Behav. 2016;6(11):e00545. Available at https://onlinelibrary.wiley.com/doi/10.1002/brb3.545.

147. Procalcitonin (PCT). Medscape, 2022. Available at https://emedicine.medscape.com/article/2096589-overview#a5 (accessed July 22, 2022).

148. Krausz AD, Korley FK, Burns MA. The current state of traumatic brain injury biomarker measurement methods. Biosensors 2021;11(9):1–22.

149. Neher MD, Keene CN, Rich MC, Moore HB, Stahel PF. Serum biomarkers for traumatic brain injury. South Med. J. 2014;107(4):248–255. Available at http://sma.org/southern-medical-journal/article/serum-biomarkers-for-traumatic-brain-injury.

150. Taghdiri F, Multani N, Ozzoude M, Tarazi A, Khodadadi M, Wennberg R, et al. Neurofilament-light in former athletes: A potential biomarker of neurodegeneration and progression. Eur. J. Neurol. 2020;27(7):1170–1177.

151. Bagnato S, Grimaldi LME, Di Raimondo G, Sant'Angelo A, Boccagni C, Virgilio V, et al. Prolonged cerebrospinal fluid neurofilament light chain increase in patients with post-traumatic disorders of consciousness. J. Neurotrauma. 2017;34(16):2475–2479. Available at http://www.liebertpub.com/doi/10.1089/neu.2016.4837.

152. Dickstein DL, De Gasperi R, Gama Sosa MA, Perez-Garcia G, Short JA, Sosa H, et al. Brain and blood biomarkers of tauopathy and neuronal injury in humans and rats with neurobehavioral syndromes following blast exposure. Mol. Psychiatry. 2021;26(10):5940–5954.

153. Gao W, Zhang Z, Lv X, Wu Q, Yan J, Mao G, et al. Neurofilament light chain level in traumatic brain injury. Medicine (Baltimore). 2020;99(38):e22363.

154. Mondragón-Rodríguez S, Salgado-Burgos H, Peña-Ortega F. Circuitry and synaptic dysfunction in Alzheimer's disease: A new tau hypothesis. Neural. Plast. 2020;2020:1–11. Available at https://www.hindawi.com/journals/np/2020/2960343/.

155. Cherry JD, Meng G, Daley S, Xia W, Svirsky S, Alvarez VE, et al. CCL2 is associated with microglia and macrophage recruitment in chronic traumatic encephalopathy. J. Neuroinflammation 2020;17(1):1–12.

156. Cherry JD, Tripodis Y, Alvarez VE, Huber B, Kiernan PT, Daneshvar DH, et al. Microglial neuroinflammation contributes to tau accumulation in chronic traumatic encephalopathy. Acta Neuropathol. Commun. 2016;4(1):112. Available at https://actaneurocomms.biomedcentral.com/articles/10.1186/s40478-016-0382-8.

157. Castellani RJ. Chronic traumatic encephalopathy: A paradigm in search of evidence? Lab. Investig. 2015;95(6):576–584. Available at http://www.nature.com/articles/labinvest201554.

158. Bezek S, Biberthaler P, Martinez-Espina I, Bogner-Flatz V. Pathophysiology and clinical implementation of traumatic brain injury biomarkers: Neuron-specific enolase. In: Wu AHB, Peacock WF. (Eds.). Biomarkers for Traumatic Brain Injury (pp. 169–182). Elsevier, 2020. Available at http://dx.doi.org/10.1016/B978-0-12-816346-7.00011-7.

159. Stålnacke BM, Tegner Y, Sojka P. Playing soccer increases serum concentrations of the biochemical markers of brain damage S-100B and neuron-specific enolase in elite players: A pilot study. Brain Inj. 2004;18(9):899–909.

Chapter 8

The Healthy Brain: More Than a Wrinkled Supercomputer

Introduction

The human brain, your personal powerhouse! It is a complex system made up of neurons, and it is so elaborate in its functioning that at its simplest level, we can relate it to the most advanced computing systems. Being made up of billions of neurons, this enables it to think deeply, feel emotions, and even make our toes wiggle (Figure 8.1). In brief, healthy brain equates to a healthy body, which further translates into longer lifespan filled with joy and fulfillment. "Brain health" until recently more or less sounded like that unused cupboard in your house that nobody talked about and yet even a clear definition was lacking. But today, science is beginning to understand the importance of "brain health." Now, science is finally catching on, and suddenly, "brain health" is the new buzzword. Brain health refers to the optimal functioning of the brain, encompassing cognitive abilities, emotional well-being, and physical health.

 DOI: 10.1201/9781032676579-8

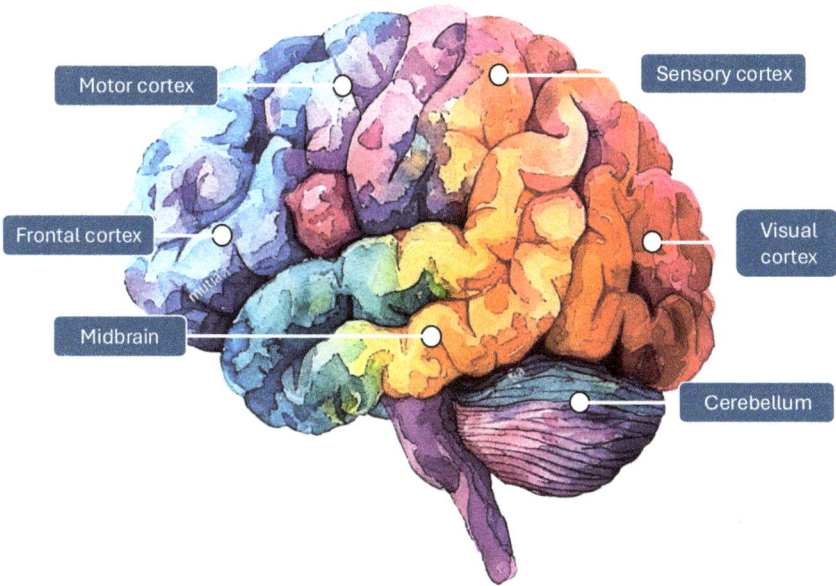

Figure 8.1   Figure of human brain.

Obtained from https://www.shutterstock.com/image-vector/brain-vector-illustration-watercolor-style-2445955681. Labeled manually using Microsoft Power point.

But it is not all moonlight and roses, quite to the contrary, in lower- and middle-income settings (for example, Pakistan), neuropathological diseases prevail. Believe it or not, these are currently the leading causes of mortality after cardiovascular malignancies resulting into 9 million body counts each year globally. The truth is your brain's well-being isn't something to mess around with.

So, what's the brain up to all day? Its activity has been broken into four main tasks by experts: perception, action, cognition, and emotion. Motor skills, cognition, emotions, hearing, and vision are the five key domains for comprehensive understanding of brain health. Inspired by the concept of the "Cognitive Clock," this approach enables us to look beyond chronological age when assessing brain functioning in individuals. In 2021, Boyle et al. pointed out that "cognitive age" counts more than how many candles you are blowing out each year when it comes to brain health and introduced such idea as "Cognitive Clock," hence an inner clock timing to show how your brain is holding up.

However, just as the Marshall Plan became a pivotal initiative after its development, brain health is an essential part of human welfare that remains often overlooked.

Imagine Europe after World War II. It was a destroyed continent with broken economies where people were starving to death every day. At that time, the United States introduced a big idea known as the Marshall Plan. It spent over billions on building Europe again; it restored infrastructures and industries while providing food and support to millions.

War-ravaged nations must have looked up at the 1948 Marshall Plan as an inviting beacon of hope, offering both resources and an encouragement for recovery. Cognitive decline, mental health issues, and neurological disorders could as well be compared with the devastation brought about by World War II. But unlike war, there is no immediate visible destruction, hence making the disorders seem slow in progression without any sense of alarm or urgency on their part.

Just as the Marshall Plan brought hope for Europe after the war, so does the investment into research, education, and support programs for brain health offer individuals a way of moving forward beyond limitations imposed by their conditions. The lack of awareness, funding, or political support makes any effort at improving brain health unlikely in many cases; yet, this neglect could mean more than just individual loss—it poses long-term dangers for societies too.

Through recognition of the importance of brain health coupled with allocation of resources appropriately commensurate investment thereon will ensure a lifelong attainment of well-being by people. In much the same way that the international community witnessed the efficacy with which the United States consolidated its power by post–World War II rehabilitation (such acts are depicted by the Marshall Plan), a global approach toward addressing concerns related to brain health would have myriad benefits for different individuals and communities worldwide.

Biomarkers: Your Brain's Report Card

Picture your thyroid, a butterfly-shaped mass in your neck, is behaving strangely. Let's presume there is a thyroid neoplasm. In this context, there are two primary choices: the thyroid scan or a biopsy. Let's find out about their differences in the simplest manner ever proposed.

Think of a thyroid scan as a tech-savvy sleuth that uses a specific substance known as radioactive iodine, which your thyroid gland has high affinity for in order to mark its territories. By tracing back the path of iodine, this scan may develop a blueprint of your thyroid and hence detect any suspicious parts. Would you like to know how it works? In fact, a thyroid scan is usually devoid of any physical injury since there are no pricks or cuts involved. Also, it takes less time to carry out since it only lasts for a duration of around 30 minutes from the time an individual swallows some radioactive iodine pill or drink. Navigating by GPS is exactly what a thyroid scan is all about—the scan shows where the doctor should go within this organ in cancer treatment process or how the person should go during treatment process.

But this isn't the case with a biopsy—it is like taking a tiny piece of the suspicious area to the microscope and seeing if it contains cancer cells or not. The gold standard for confirming if a tumor is indeed benign or malignant is the biopsy procedure. These results help doctors decide what steps to take next. So, which one is better? Many times, doctors prefer doing a thyroid scan first so that they know generally what is going on.

If something abnormal comes out on the scan, the biopsy procedure provides a definitive diagnosis. Imagine you're using a map prepared by a detective to search for some clue, then you resort to a magnifying glass to make observations.

Monitoring brain health is far from easy. Ideally, there should be some magical test for measuring brain functions—like checking your cholesterol levels—but for the head or brain. That's where biomarkers come into play. These biomarkers are like tiny messengers that float around in our bloodstreams and give us clues about what's going on inside our heads.

However, despite over half-a-century of researches, scientists are still questioning themselves as to which approach would be most suitable when it comes to assessing how well our brains work. Instruments that can provide accurate assessments while being affordable and user-friendly are needed. What's more, they must be minimally invasive; wouldn't you agree that nobody wants an unbearable brain test or perhaps one feels like torture at any point?

The problem? No one can agree on a standard way to define brain health, and that's making it tough to include this in global health programs. But with the world's population getting older and mental health concerns rising faster than ever, we need to figure this out fast.

Imagine you are at work one day and your boss calls you into his office, he tells you that he is concerned about your performance. You may have been working well for some time, but recently things seem to have changed. The problem starts when you stop working instead start doing something completely different. This problem aggravates when you find yourself in unfamiliar places such as the supermarket without even knowing how you got there.

The above scenario is of symptoms displayed by sufferers of Alzheimer's disease, who randomly recall events and places experienced in childhood but are unable to remember what they did yesterday afternoon or even this morning without traces left behind. Sometimes these people choose words carefully and respond slowly when asked simple questions, yet more often than not they speak loudly and speak rubbish at times using the wrong language altogether. Disease tends to sneak up on them silently like a thief in the night, before finally destroying any meaning attached to their lives.

But let's face it: Despite years of research, scientists still don't know how to measure brain health in the best way. The tools we need must be accurate, affordable, and user-friendly. This is why we also need them to be less painful because no one wants to go through a torture when taking a test.

The Global Brain Health Institute: Your Brain Defense

Introducing the Global Brain Health Institute—an initiative dedicated to addressing this intellectual enigma. It works toward enhancing the awareness of brain health globally, and especially so in regions like Pakistan, where there is an acute lack of knowledge on how best one can assess brain function affordably.

This is where the fun lies—predictive modeling. Sounds like some complex stuff, huh? Well, it is. Whether diagnosing diseases, predicting treatment outcomes, or determining its long-term effects, predictive modeling comes in handy. Basically, it serves as a crystal ball for brain health. Imagine predictive modeling as a crystal ball, but instead of looking ahead, it delves into the past and present to draw conclusions on future events.

Just as fortune tellers rely upon their knowledge of previous happenings and tendencies to make predictions, predictive models employ historical data analysis as well as trend identification; hence can deduce information about what is about to happen.

Cracking the Code: A Brain Health Formula

So, what was my central research objective? To finally come up with a formula for assessing the health of one's brain—a template through which it is possible to determine how well someone's brain functions, by evaluating its key faculties and using biomarkers as indicators.

In pursuit of these objectives, this study complied with all rules (including STROBE and TRIPOD); this study was conducted in Peshawar, Pakistan, over a span of one year. In this period, researchers collected 75 subjects (healthy individuals aged between 19 years old and 35 years old) through advertisements on social media platforms, posters, or by word of mouth. Participants were selected based on certain features: For example, they should not have heart diseases nor any kidney or liver problems; there should be absence of psychiatric disorders as well as heavy coffee intake.

Afterward appropriate people were examined for seven separate abilities affecting the brain such as vision, hearing capacity, cognition levels, or general thinking ability with emphasis upon memory retention. Balance, coordination, and emotions were also measured. Then, we came up with a brain health formula through the magic of multiple linear regression (but don't worry—no math is happening here).

Vision, Grip, and All That Jazz: How the Tests Were Done

For vision, subjects had their visual fields checked using a student perimeter. Basically, you stare straight ahead while someone checks if you can still see what's happening on the sides—that's essentially what the test is about. If you could still spot things happening all around you, great! If not, you got a score based on how much vision you have lost.

After that came the hand grip strength test which involved getting participants to squeeze the dynamometer as hard as they could manage; think of it like the ultimate handshake challenge. We took three readings each time before averaging their final result for each individual.

Moving on to hearing, individuals had to put on earphones after which an audiometer was used in order to test their auditory abilities depending upon whether they could hear anything or not at all. This was further classified into normal hearing, mild hearing impairment, moderate hearing loss, or severe hearing deficit based on residual auditory function.

For emotions, there was a 30-item questionnaire called TEIQue-SF that asked participants to rate statements from 1 (completely disagree) up to 7 (completely agree). Meanwhile, cognitive function was measured using the MMSE—a test with 30 questions that see how you think or remember stuff or solve problems.

Now let us talk about biomarkers! NSE, NfL, and pTau are some proteins present in our brains and we use a technique called the ELISA to determine their concentrations; this helps us know if we are at risk of developing dementia or other brain disorders in our old age (Figure 8.2).

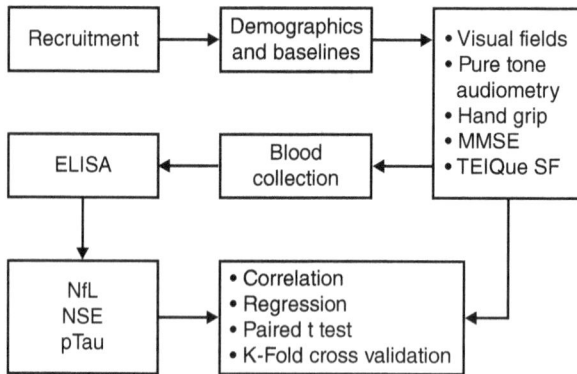

Figure 8.2 Flowchart showing methodology of Phase 3.

Results: The Brain Report Card

Of the 75 participants, men to women ratio was equal. What a relief that most of them had normal sight and hearing abilities! Specifically, it was established that 81.3% displayed normal visual fields, just as hearing in 78.7% was like a bat's jealousy. On the other hand, an average of 30.43 kg was the dominant hand-squeeze force during grip strength testing!

This time round, the story changed when it came to things concerning intelligence as registered by an average score of 144.72 in terms of emotional wisdom on the one hand. On the other hand, cognition was under control through a mean MMSE value of 29.97 on average across all individuals who took part in the study. As for the biomarker levels traveling from their brain to their blood, NfL was 0.026 ng/ml, NSE was 10.46 ng/ml, and pTau was 4.48 ng/ml (Figure 8.3).

Can We Predict Brain Health?

It is at this point where things start getting interesting. The result showed that for NfL levels, only emotional intelligence measured by the TEIQue-SF stood out as a significant factor

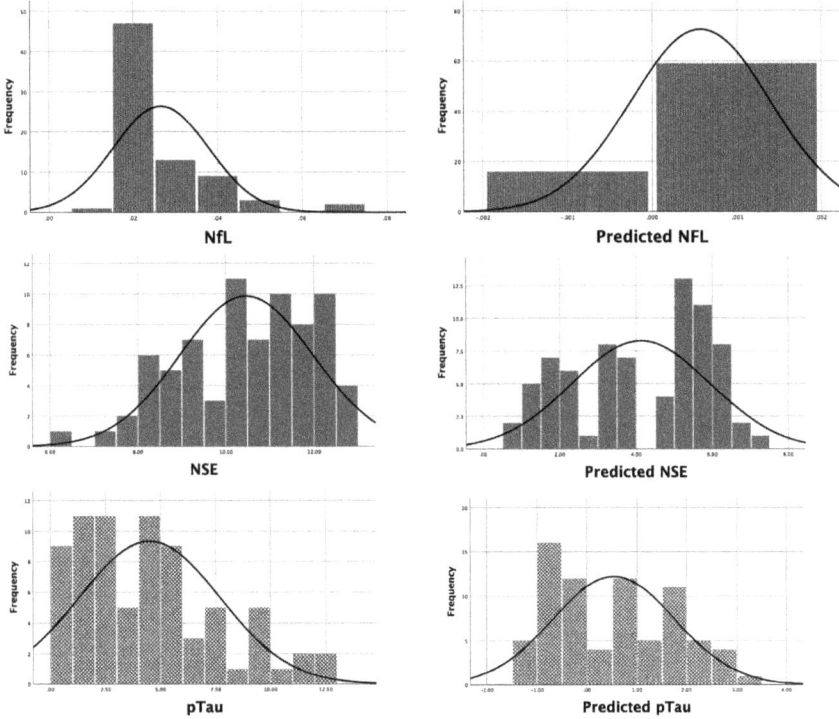

Figure 8.3   Histograms comparing actual and predicted neuro-biomarker levels in the participants of the study.

when comparing univariate regression models to multivariate ones. Simplifying it further, if you are good at keeping your emotions in check, it could be that things are alright with your mind.

With pTau, which is a biomarker associated with diseases such as Alzheimer's, both age and handgrip strength were significant predictors for univariable analysis. Nevertheless, when multiple factors were taken together, emotional intelligence came up once again as significant determinant.

In conclusion, researchers said healthy brain behaves just like a well-oiled machinery would do; trying to predict brain health from individual components alone tends not to be useful because its different parts function so harmoniously than it cannot be predicted as a whole. Thus, it is like evaluating the health of an individual by simply looking at his general physical demeanor without doing any specific tests to check for a specific condition affecting a particular organ or organ system.

Limitations and Final Thoughts

Now, before you get too excited about these findings, let's talk limitations. For one, the study didn't use advanced brain imaging techniques like EEG, PET, or fMRI scans, which could have provided more detailed insights.

That being said, the study did manage to come up with a cost-effective and feasible way to assess brain health. And in a world where health budgets are often tight, that's no small feat.

Conclusion: The Brain Formula That Almost Was

In conclusion, the formula devised to measure brain health was a step in the right direction, but it fell short of being a foolproof predictor. Why? Because healthy brains are complex, harmonious systems, and their parts work together too smoothly to break them down into simple equations. So, while this study didn't crack the code on predicting brain health entirely, it did shed light on how different brain functions work together to keep us going strong (Figure 8.4).

So, the next time you're sipping your morning coffee and feeling that brainpower kick in, give your noggin a little thanks—after all, it is doing a lot more than just processing your latest Instagram feed.

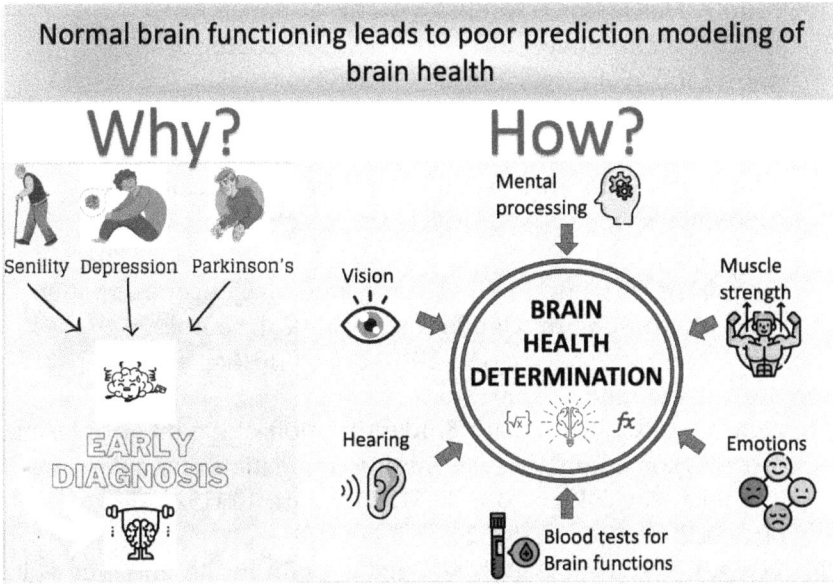

Figure 8.4   The figure depicts how early diagnosis of brain disorders is mandatory for maintenance of brain health.

Bibliography

1. What is a healthy brain? New research explores perceptions of cognitive health among diverse older adults. 2007.
2. What if mental health was brain health? 2017.
3. Brain health. 2022. WHO. file:///Users/umemazafar/Desktop/Thesis PhD/Intro chap 1/Brain health WHO.html.
4. Nall R. Brain damage: What you need to know. 2021. Healthline. Available at https://www.healthlineeffects of traumatic brain injury (TBI)? https://www.nichd.nih.gov/health/topics/tbi/conditioninfo/effects#.
5. Lloyd D. Scanning the neurocracy: What do Brodmann areas do? Lloydia 1987;11:23–78. Available at http://commons.trincoll.edu/dlloyd/research/videos-web-pages/scanning-the-neurocracy-what-do-brodmann-areas-do/.
6. Krausz AD, Korley FK, Burns MA. The current state of traumatic brain injury biomarker measurement methods. Biosensors 2021;11:1–22. doi: 10.3390/bios11090319.

7. Todd K Shackelford, Weekes Viviana A. Functional specialization. Encycl. Evol. Psychol. Sci. 2021;3297–3297. doi: 10.1007/978-3-319-19650-3_301963.

8. Boyle PA, Wang T, Yu L, Wilson RS, Dawe R, Arfanakis K, Schneider JA, Beck T, Rajan KB, Evans D, et al. The "cognitive clock": A novel indicator of brain health. Alzheimers Dement. 2021;17(12):1923–1937. doi: 10.1002/alz.12351.

9. Zetterberg H, Blennow K. Chronic traumatic encephalopathy: Fluid biomarkers. In: Handbook of Clinical Neurology (pp. 323–333). Elsevier B.V., 2018. doi: 10.1016/B978-0-444-63954-7.00030-6.

10. Zafar U, Habib SH, Bashir S. Identification of a panel of plasma protein biomarkers for determining traumatic encephalopathy in adults. Physiology 2023;38:5728352. doi: 10.1152/physiol.2023.38.S1.5728352.

11. Wang Y, Pan Y, Li H. (2020). What is brain health and why is it important? BMJ 371, m3683. doi: 10.1136/bmj.m3683.

12. Manwell LA, Barbic SP, Roberts K, Durisko Z, Lee C, Ware E, McKenzie K. What is mental health? Evidence towards a new definition from a mixed methods multidisciplinary international survey. BMJ Open. 2015;5:e007079–e007079. doi: 10.1136/bmjopen-2014-007079.

13. Khanna R, Edgcomb JB. Brain health diplomacy in a new decade: Why innovation matters. Int. Psychogeriatrics 2020;32:901–903. doi: 10.1017/S1041610220000381.

14. Poldrack RA, Huckins G, Varoquaux G. Establishment of best practices for evidence for prediction. JAMA Psychiatry 2020;77:534. doi: 10.1001/jamapsychiatry.2019.3671.

15. Collins GS, Reitsma JB, Altman DG, Moons KGM. Transparent reporting of a multivariable prediction model for individual prognosis or diagnosis (TRIPOD): The TRIPOD statement. Ann. Intern. Med. 2015;162:55–63. doi: 10.7326/M14-0697.

16. Nau R. What's a good value for R-squared? Stat. Forecast. Notes Regres. Time Ser. Anal. 2020;1–10. Available at https://people.duke.edu/~rnau/rsquared.htm%0A https://people.duke.edu/~rnau/rsquared.htm%0A https://people.duke.edu/~rnau/rsquared.htm%0A http://people.duke.edu/~rnau/rsquared.htm.

17. Chen Y, Demnitz N, Yamamoto S, Yaffe K, Lawlor B, Leroi I. Defining brain health: A concept analysis. Int. J. Geriatr. Psychiatry 2022;37. doi: 10.1002/gps.5564.

18. Başar E. The theory of the whole-brain-work. Int. J. Psychophysiol. 2006;60:133–138. doi: 10.1016/j.ijpsycho.2005.12.007.

19. Badcock PB, Davey CG, Whittle S, Allen NB, Friston KJ. The depressed brain: An evolutionary systems theory. Trends Cogn. Sci. 2017;21:182–194. doi: 10.1016/j.tics.2017.01.005.

20. Raichle ME. The restless brain: How intrinsic activity organizes brain function. Philos. Trans. R. Soc. B Biol. Sci. 2015;370:20140172. doi: 10.1098/rstb.2014.0172.

21. Price CJ. A review and synthesis of the first 20 years of PET and fMRI studies of heard speech, spoken language and reading. Neuroimage 2012;62:816–847. doi: 10.1016/j.neuroimage.2012.04.062.

22. Vimont C. What does 20/20 vision mean? Am. Acad. Ophthalmol, 2023. Available at https://www.aao.org/eye-health/tips-prevention/what-does-20-20-vision-mean.

24. InformedHealth.org. Hearing loss and deafness: Normal hearing and impaired hearing. NCBI Bookshelf, National Library of Medicine (Institute for Quality and Efficiency in Health Care), 2017.

25. How to identify the most important predictor variables in regression models. Minitab, 2016, pp. 1–6. Available at https://blog.minitab.com/en/adventures-in-statistics-2/how-to-identify-the-most-important-predictor-variables-in-regression-models.

25. Lau ZJ, Pham T, Chen SHA, Makowski D. Brain entropy, fractal dimensions and predictability: A review of complexity measures for EEG in healthy and neuropsychiatric populations. Eur. J. Neurosci. 2022;56:5047–5069. doi: 10.1111/ejn.15800.

26. Yang AC, Tsai S-J. Is mental illness complex? From behavior to brain. Prog. Neuro-Psychopharmacol. Biol. Psychiatry 2013;45:253–257. doi: 10.1016/j.pnpbp.2012.09.015.

27. Paulus MP, Geyer MA, Braff DL. Use of methods from chaos theory to quantify a fundamental dysfunction in the behavioral organization of schizophrenic patients. Am. J. Psychiatry 1996;153:714–717. doi: 10.1176/ajp.153.5.714.

28. Vaillancourt DE, Newell KM. Changing complexity in human behavior and physiology through aging and disease. Neurobiol. Aging 2002;23:1–11. doi: 10.1016/S0197-4580(01)00247-0.

Chapter 9

Unlocking the Diseased Brain's Secrets

Introduction

The human brain—having a jelly-like texture—is responsible for coordinating everything that we do, think about, or feel. Yet, despite being the most important organ in our body, brain health has often been neglected in medical discussions. Attempting to explain schizophrenia or major depression without looking at the brain is like baking cake with only water and sugar and skipping out flour—it will not come out well! Unfortunately, this has often been sidelined from broader medical forum without adequate understanding among practitioners who sometimes have limited knowledge base regarding these conditions relative to their own areas of specialization.

The more we understand about mental illnesses, the clearer they become. The brain and the body are not separate entities, they are like two sides of a coin—they cannot exist separately. For example, it is difficult to manage schizophrenia or major depression without taking the brain's health into

 DOI: 10.1201/9781032676579-9

account. Merely symptomatic treatment won't do! The root cause of disorders like these needs to be nibbed in the bud. Hence, before even thinking of eliminating such illnesses, the brain's anatomical and physiological health must be taken into account. This is so because when you are sick with a bad headache or any other problem associated with the brain, the whole body system is at risk of collapse.

But here's the thing—certain illnesses such as multiple sclerosis, stroke, Alzheimer's, Meniere's disease, or amblyopia normally begin in a small unnoticed manner before they can snowball into large neurological nightmares. The challenge is that diagnosing brain diseases always involves some form of painful tests like spinal tapping or exposure to radiation. Ouch! Clearly, a better solution is needed.

The Application of Biomarkers

That's where biomarkers like neurofilament light chain (NfL), neuron-specific enolase (NSE), and phosphorylated Tau (pTau) come in. Simply put, these three function as your brain's warning signs. If things start going wrong, these little markers spike and tell us something's up. This could help in managing diseases before they encroach upon your mind like an unwanted guest who stays too long.

Putting Brain Health under the Microscope: How Does It All Work?

Imagine your brain as an automobile. You need to make all its components work so that it runs smoothly. Components like the radiator, engine, and brakes should be up and running for the car to perform its core function—transport. Similarly, vision, hearing, and cognition need to be apt for the brain to work unhindered. Now, imagine there was a bodily function dashboard that could give information on how healthy your brain was depending on the operations of each system that it controls, would not it be nice?

This research study created formulas that combine brain function tests and biomarker levels to give an idea of overall brain health. For instance, this approach can be used by clinicians as a way of determining if the patients' brains are alright or if they might have difficulty down the road. And the best part is that this method was noninvasive, cost-effective, and reliable.

Participants who suffered from various diseases that affect the brain, including amblyopia (lazy eye), stroke, Alzheimer's disease, Meniere's disease, depression, and senility, were recruited through flyers, social media, or hospital visits at the outpatient departments (Figure 9.1).

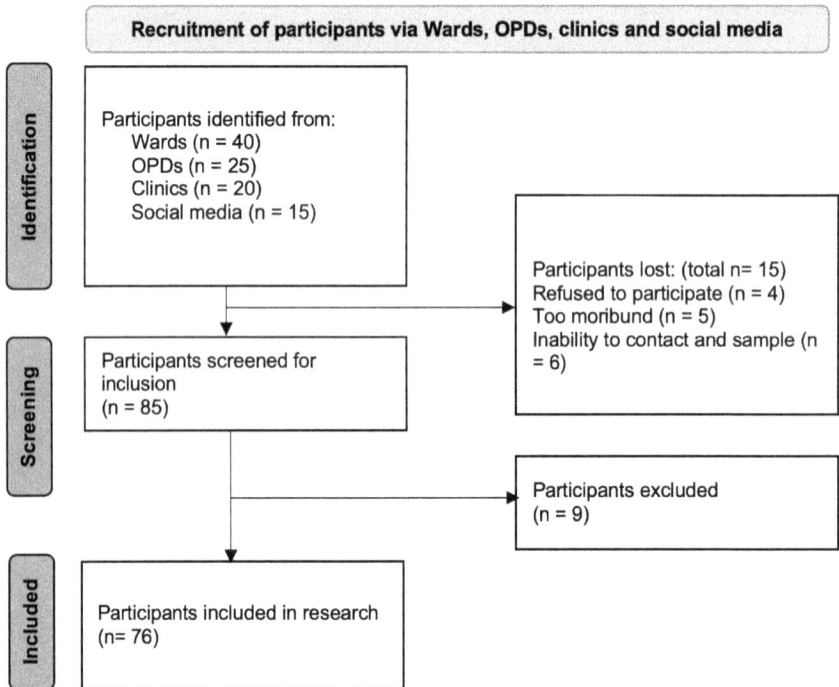

Figure 9.1 Flowchart showing participant recruitment.

These tests evaluated seven major brain functions—vision, hearing, cognition, strength, balance, emotions, and age—while three key biomarkers were profiled, namely, NfL (neurofilament light chain), NSE (neuron-specific enolase), and pTau (phosphorylated Tau). These biomarkers act like report cards for the brain and show how it is doing under disease conditions.

The Biomarker Trio: NfL, NSE, and pTau—Your Brain's Early Warning System

Let's break down these three biomarkers and what they tell us about brain health (Figure 9.2):

■ NfL (neurofilament light chain) is used as an indicator showing any form of damage done to neurons within one's body, thus playing similar role in stress response mechanisms present during bodily traumatic conditions. High levels of NfL in the blood are a sign that axons (long fibers carrying signals between neurons) are getting damaged. The presence of high quantities NfL bloodstream indicates that something is wrong somewhere, mainly occurring in instances like stroke, where cells die rapidly.

■ NSE (neuron-specific enolase) indicates when nerves have been damaged or are under stress. This is the kind of inspector that keeps an eye on the neurons, telling us if they are in trouble. It is used more for head traumas, but it can as well be critical in cases of strokes or old age syndromes characterized by neuronal loss.

■ pTau (phosphorylated Tau), in addition to forming tangles inside the nerve cells, plays other vital roles in various neurological disorders like Meniere's syndrome and depression leading to decline in mental abilities such as

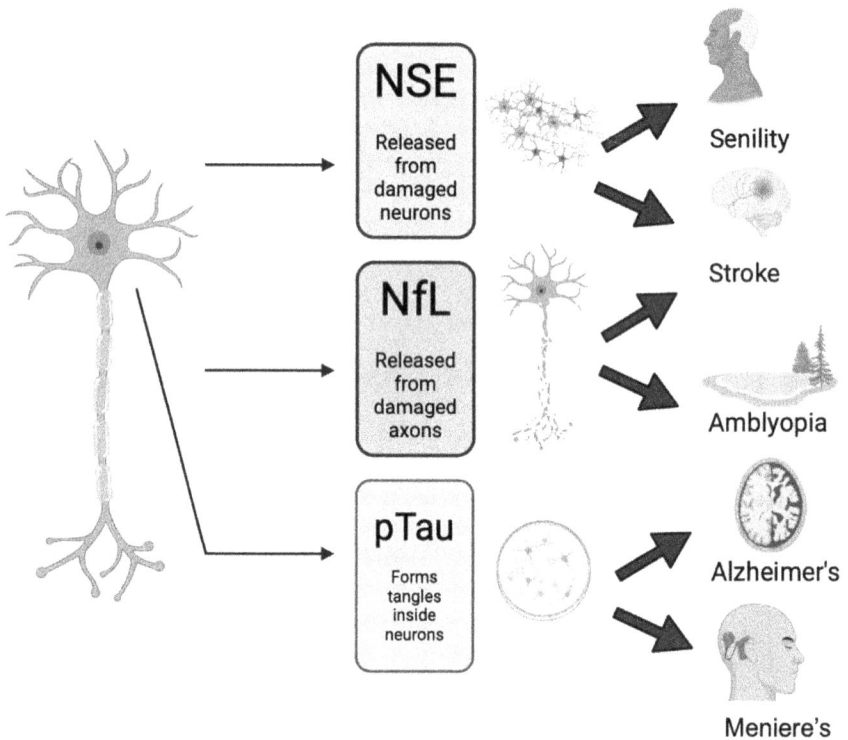

Figure 9.2 Flowchart designed using biorender illustrating how damage to neurons leads to the release of specific biomarkers.

memory loss or poor judgment. This helps monitor the brain's overall health and its ability to cope with damage.

An Extensive Examination of Brain Health in Different Illnesses

Now, let's dig into the nitty-gritty of what the study found. Among others, all participants were subjected to a series of tests aimed at evaluating their brain function from sight and sound to emotions and equilibrium. Furthermore, the scientists went ahead to determine how each participant's levels of NfL, NSE, and pTau concurred with brain activity.

Here is what was found:

Amblyopia: The Lazy Eye That's Not So Lazy

When you hear about amblyopia (or "lazy eye"), it sounds like something related to the eyes only but there is much more than that: a deeper relationship with the state of the brain. Results also showed that improvements in vision were associated with increased levels of NfL. This is consistent because NfL helps in stabilizing neurons, hence supporting brain plasticity, which allows the brain to adapt and recover. In simpler terms, the possibility that NfL might be assisting in self-repair mechanisms of the brain does not seem far-fetched even among adult sufferers of amblyopia since this condition was believed to be irreversible once out of childhood.

Alzheimer's Disease: Tracking Cognitive Decline

Alzheimer's is the big bad wolf of brain diseases. Scores for MMSE (Mini-Mental State Exam), which directly measures how your mind is working, have been found to be directly related to the levels of NfLs here during the investigation. When cognition gets worse, higher levels of NfL are present—it feels like your mind is having difficulty keeping up with life in this context. This link between NfLs and cognitive abilities is important for understanding what leads to Alzheimer's disease, as doctors may then track the effectiveness of treatments over time.

Meniere's Disease: The Ear–Brain Connection

Meniere's disease primarily causes hearing impairment and vertigo; however, this study also demonstrated its link with pTau levels. pTau is usually associated with Alzheimer's and it seems to perform one of those cleanup roles for brains which need help after being damaged. Basically, think of pTau as one big cleanup crew for the brain, cleaning out debris and aiding in a speedier recovery process by the nerve cells around which they are produced. A correlation between high pTau levels was found in Meniere's patients with favorable results indicating that this might be a good marker for disease progression monitoring.

Depression: When Emotions Affect the Brain

Usually, we consider depression primarily in terms of emotions; however, this pathology is a brain disorder as well. Prolonged depression destroys your normal way of functioning, leading to cognitive impairment or even brain injury. This study found that pTau levels correlate with both cognition status and mood. (Figure 9.3) What it means is

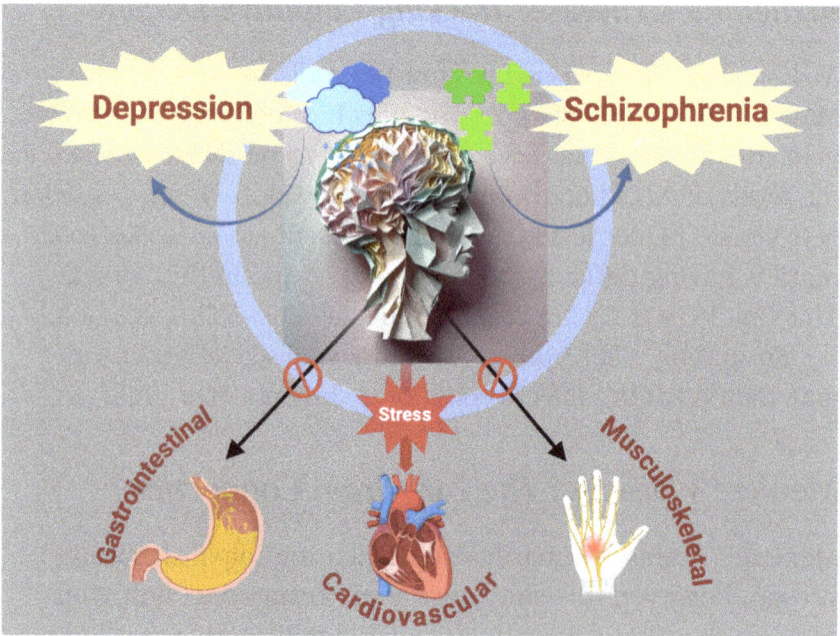

Figure 9.3   Effect of depression and schizophrenia on various body systems.

that as depression worsens, the pTau amounts rise, further implying that in the end depression truly harms our brains. And if you thought depression was just in your head, well, this study proves it is actually in your brain too!

Stroke: A Full-On Brain Assault

When stroke occurs, it is as if a storm has hit your brain. This is characterized by blood deprivation and cell death leading to massive damage. The key markers of brain injury in patients with stroke were elevated levels of all three biomarkers: NfLs, NSEs, and pTaus. The persistent elevation of these indicators beyond the initial stroke period is noteworthy as they serve to aid physicians document progress toward recovery. Especially regarding the healing process, NfL plays a significant role here; it tells us if our brain is doing well with respect to cure or not, in other words, if it continues struggling.

Senility: When Age Takes Its Toll

Finally, in the senile population, NSE was the most telling biomarker. Furthermore, NSE levels increase with age due to the loss of neurons. NSE has been found to be very significant in tracking neuronal health among old people and evaluation of how much brain function has been lost as a result of advancing years. Rather, it stands for an index of "wear and tear" of the brain in individuals who continue to age healthily.

The Big Reveal: Can We Predict Brain Health?

A step ahead of just gathering information was coming up with real formulas that could foretell one's well-being based on these biological markers' concentration in blood combined with specific brain function results (or scores). These models were specific for a specific illness group, but the findings were crystal clear that pTau was the most dependable predictor of brain health in these different sets of diseases. This is what we found out:

pTau-based formulas were very effective predictors of brain health particularly in conditions such as Alzheimer's disease, strokes, and depression, because it indicates the degeneration of neurons and hence plays a crucial role in assessing brain damage in different diseases.

By contrast, the NfL-based formulas seem to perform better in cases like stroke, where levels of NfL stay high several months after the attack. This makes NfL an appropriate tool for checking how well someone recovers from a stroke as well as monitoring for long-term recovery processes.

Besides senility and depression, NSE-based formulas proved to be useful, especially when the latter is caused by damage to nerve cells.

What This Portends for Brain Health

The findings from this research have led to a paradigm shift in how we look at brain health issues. The combination of biomarker levels with tests for brain function will enable medical practitioners to have a more detailed understanding of the state of the brain at the early stages of disease progression. This means that instead of waiting for symptoms to become severe, doctors can catch brain damage early and intervene before things get worse (Figure 9.4).

For the findings of this study, the future involves the possibility of early detection and customized intervention, thanks to generation of formulas. Early identification could mean monitoring noninvasively; this might be done without waiting for the worst to happen. This news is not only reassuring but also revolutionary in terms of better brain health globally.

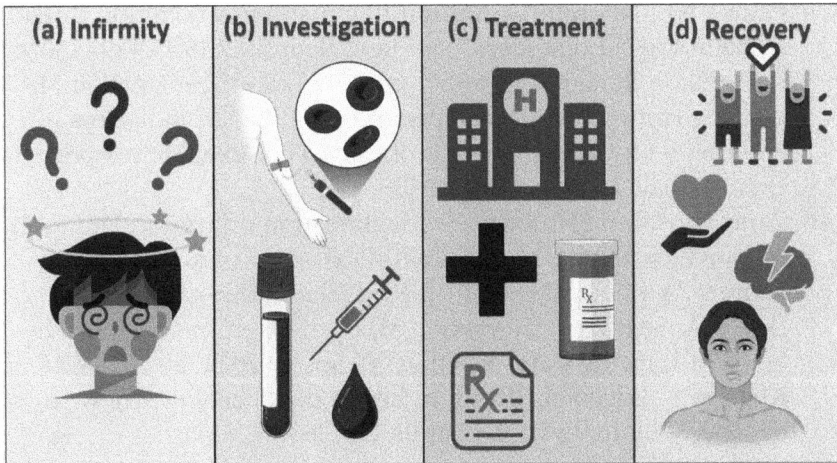

Figure 9.4   Process of progression from disease to recovery.

Then again, we should start hoping for a world where instead of necessarily trying to deal with brain damages after they have happened, there is early identification followed by monitoring as well healing from within the brain itself. And that's not just good news for patients—it is a game changer for global health.

So, here's to a future where we don't just react to brain damage—we catch it early, monitor it, and help the brain heal itself. Because at the end of the day, your brain's got a lot more going on than just reminding you where you left your keys.

Bibliography

1. Boorse C. What a theory of mental health should be. J. Theory. Soc. Behav. 1976;6:61–84.
2. Giovannoni G, Butzkueven H, Dhib-Jalbut S, et al. Brain health: Time matters in multiple sclerosis. Mult. Scler. Relat. Disord. 2016;9:S5–S48.
3. Walton EL. Saliva biomarkers in neurological disorders: A "spitting image" of brain health? Biomed. J. 2018;41:59–62.

4. Gilhus NE, Deuschl G. Neuroinflammation: A common thread in neurological disorders. Nat. Rev. Neurol. 2019;15:429–430.
5. Adelantado-Renau M, Esteban-Cornejo I, Rodriguez-Ayllon M, et al. Inflammatory biomarkers and brain health indicators in children with overweight and obesity: The ActiveBrains project. Brain Behav. Immun. 2019;81:588–597.
6. Zafar U, Habib SH, Raza SS. Identification of appropriate tools to gauge brain functions in a clinical setup of a developing country: A pilot study. Pak. J. Med. Sci. 2023;39:1840–1846.
7. Eyre HA, Searfoss J, Hopwood M, et al. Open innovation: The key to advancing brain health. CNS Spectr 2022; 28:392–394.
8. Khanna R, Edgcomb JB. Brain health diplomacy in a new decade: Why innovation matters. Int. Psychogeriatrics 2020;32:901–903.
9. Pekny M, Wilhelmsson U, Stokowska A, et al. Neurofilament light chain (NfL) in blood: A biomarker predicting unfavourable outcome in the acute phase and improvement in the late phase after stroke. Cells 2021;10:1537.
10. Sengpiel F. Amblyopia: Out of the dark, into the light. Curr. Biol. 2013;23:R195–R196.
11. Duffy KR, Mitchell DE. Darkness alters maturation of visual cortex and promotes fast recovery from monocular deprivation. Curr. Biol. 2013;23:382–386.
12. Holman KD, Duffy KR, Mitchell DE. Short periods of darkness fail to restore visual or neural plasticity in adult cats. Vis. Neurosci. 2018;35:E002.
13. Reddy PH, Oliver DM. Amyloid beta and phosphorylated tau-induced defective autophagy and mitophagy in Alzheimer's disease. Cells 2019;8:488.
14. Mahaman YAR, Embaye KS, Huang F, et al. Biomarkers used in Alzheimer's disease diagnosis, treatment, and prevention. Ageing Res. Rev. 2022;74:101544.
15. Botto R, Callai N, Cermelli A, et al. Anxiety and depression in Alzheimer's disease: A systematic review of pathogenetic mechanisms and relation to cognitive decline. Neurol. Sci. 2022;43:4107–4124.
16. Banning LCP, Ramakers IHGB, Deckers K, et al. Affective symptoms and AT(N) biomarkers in mild cognitive impairment and Alzheimer's disease: A systematic literature review. Neurosci. Biobehav. Rev. 2019;107:346–359.

17. Babulal GM, Ghoshal N, Head D, et al. Mood changes in cognitively normal older adults are linked to Alzheimer's disease biomarker levels. Am. J. Geriatr. Psychiatry 2016;24:1095–1104.
18. Hoffmann J, Janowitz D, Van der Auwera S, et al. Association between serum neuron-specific enolase, age, overweight, and structural MRI patterns in 901 subjects. Transl. Psychiatry 2017;7:1272.
19. Papuć E, Rejdak K. Increased cerebrospinal fluid S100B and NSE reflect neuronal and glial damage in Parkinson's disease. Front. Aging Neurosci. 2020;12:1–8.
20. Toledo ARL, Monroy GR, Salazar FE, et al. Gut–brain axis as a pathological and therapeutic target for neurodegenerative disorders. Int. J. Mol. Sci. 2022;23:1184.
21. How to identify the most important predictor variables in regression models. Minitab, 2016; Accessing date: 6th June 2024. https://blog.minitab.com/en/adventures-in-statistics-2/how-to-identify-the-most-important-predictor-variables-in-regression-models.
22. Salkind N. Encyclopedia of Research Design. Thousand Oaks, CA: SAGE Publications, Inc., 2010. doi: 10.4135/9781412961288.

Chapter 10

Cracking the Code of Brain Health

Introduction

It has been established in the previous chapters that brain health can be a tricky thing. Many may argue that our minds are something of a wonder of the world, always working day and night. What the brain demands is no different from how in most cases an individual expects for her car that has been driven for some distances and on several occasions needs to be serviced. However, it's not always this simple.

For years, scientists have been trying to measure brain health from every possible angle—and trust me, that's no small task. But here's the exciting part: research is finally filling that giant gap in our understanding. And I'm here to walk you through it. We have hit an exciting milestone; long-awaited research results are being realized, and I am prepared to explain this to you in a more interesting and fun manner.

This was not a quick fix strategy. Four distinct and well-planned "phases" took place before this research reached its present state. First things first? Boiling down a laundry list

 DOI: 10.1201/9781032676579-10

of brain functions into five key categories. We rummaged through the brain's complete inventory to source the instruments that can adequately measure these functions. Phase two, an equivalent of a treasure hunt in the context of looking for biomarkers—those small SOS signals that denote if your brain has experienced significant stress. During the last two phases, all the aforementioned brain functions as well as chronological age of the participants were estimated, including a multiple linear regression equation. And the predictions came out right after the numbers were crunched—really knitting patterns. It was like hitting a bullseye every time.

The Standard: All You Need to Know about Healthy Brain Behavior

One may ask how to determine an individual's brain health? And what has always worked is developing a certain criterion. Assume that you are creating a specific smoothie, but rather than bananas and strawberries, it is a mix of brain functions and biomarkers. Some key ingredients can be vision, hearing, cognition, neurofilament light chain (NfL), phosphorylated Tau (pTau), and neuron-specific enolase (NSE). Mix it together in a blender, and there it is: A user-friendly technique to evaluate brain health in healthy adults which is noninvasive and surprisingly turns out to be very effective. And it achieved what it was set out to. The predictions coincided with the actual biomarker levels, confirming this model is workable.

But again, why this concern? Well, as the world's population ages, poor mental health is becoming a global public health crisis. And in countries like Pakistan, where resources are tight, there's a pressing need for cost-effective and simple ways to measure brain health. The formula we developed isn't just about predicting numbers—it's about empowering people and providing healthcare professionals with tools that can make a real difference.

Now, let's shift our focus to how this formula can be applied in different groups of patients. In regard of a partial vision loss in one eye and stroke patients, equations which use NfL and pTau as factors emerge to be quite correct. This makes sense because such two factors are indeed warning signs demanding an immediate solution in the unfortunate event when there is a disruption of the brain's nerve fibers. Episodes of depression reached peak among the study cases of patients with Alzheimer's when measuring pTau levels—we do see some smoke before the blaze. On the other hand, for geriatric patients, it would be NSE since, let's face it, as the years go by, the brain janitor (NSE) is overly used in facilitating the flow of things within the body.

A Closer Look at Our Star Biomarkers

Now, why do we need to know about these biomarkers? Let's have an insight into this with a touch of comedy and plenty of science:

> *Neurofilament Light Chain (NfL):* Let us imagine that your brain is similar to a modern office. Axon is just a telephone connecting all the departments. When these phone lines get damaged—maybe someone tripped over them—NfL gets released into the bloodstream. It's like a distress call: "Hey, we've got a problem here!" When the levels of NfL are increased, we have a good reason to think that the axons are damaged, and in that way, NfL is suitable for detecting brain injuries where there is extreme fatigue and chronic traumatic encephalopathy (CTE). Even a blood test can detect NfL at appreciable levels of sensitivity; it is, consequently, effective for routine evaluation.

Phosphorylated Tau (pTau): Let me begin by explaining what is Tau as it confuses a lot of people. Neurons are nerve cells and there are axons between two neurons. In normal circumstances Tau is a kind of scaffolding that helps to cushion the neurons. It does not cause any damages in this case. There comes a time, however, when some conditions can incite the production of more Tau in the system. With prolonged inflammation within the system or after repetitive head injuries, Tau accumulates and becomes phosphorylated or "overactive." It is comparable to workers at a construction site using an unnecessary amount of piping or tubing for cup and holder creation. Ultimately, it is bound to be a mess. Hence the name, pTau. These tangled proteins, at the end of the day, clog up the brain, resulting in issues such as dementia. When there are elevated pTau levels in the blood, it is equivalent to that entire building which has scaffolds erected in every corner on the construction site—you know something's up. This sign leads to the notion that there is some pTau in the blood and the person may be suffering from Alzheimer's or other forms of dementia.

Neuron-Specific Enolase (NSE): Picture NSE as the handyman of the brain. It swoops in to help the neurons repair themselves post damage and ensures that the body continues working by balancing the amount of glucose being produced in the body. When a hang up arises, for example, brain inflammation or injury spikes in NSE levels portray that the handyman is on his feet undertaking other chores of fixing the malfunctions. In older people, it is particularly suitable since as we grow old, the brain requires more maintenance, and in this case, NSE makes sure that the maintenance is done.

Putting It into Practice

Let's now focus on the application of the derived formulae. In a perfectly normal scenario, the brain always remains stable and reliable—just like a clock. However, the introduction of pathologies like Alzheimer's, depression, or even stroke give a new meaning to the state of the brain functions in most people. All these disorders and diseases may lead to significant disturbances of the brain functions of any organism. Despite the chaos, our formula did a pretty good job of predicting brain health, especially in the diseased populations.

For individuals with Alzheimer's and depression, pTau was most predictive of the brain health. It's like having a thermometer for inflammation—if pTau levels are high, you know the brain is struggling. In terms of stroke victims, the use of NfL and pTau was predominant with respect to the level of damage caused to the nerve fibers and subsequent damage from inflammation. There was no doubt as to who was the MVP for our elderly patients; NSE turned out to be very reliable and saved the day when it stepped in to demonstrate the efficiency of the caretaker's efforts.

Weaknesses, Challenges, and Real-World Problems

Naturally, it is impossible to create an ideal study and this one had its share of hurdles as well. Diverse logistical and operational issues arose, ranging from equipment-related technical failures, to locating the physical sites and clinical assessors, to the participants' availability. That's why we focused on affordable, easy-to-use tools that could be implemented in low- and middle-income countries. It's about making a difference where it's needed most.

Second, we focused on the suitability of the mathematical model—what can be done to avoid inherent error in the model when data about exercise, nutrition, or stress is acquired. Although it helped to concentrate on brain functions in isolation, there is another paradox: Real-life factors do impact the brain health. At times, neglecting these dynamics can skew the results, but it was a trade-off necessary for the sake of the research.

Summing up, this research sets new standards for the management of brain functions and is tailored to the needs of individuals. It might be striving to identify the early signs of Alzheimer's, evaluating changes after a stroke, or watching the deterioration during cognitive functions in the elderly age— these summed facts now offer a new means in combating this condition. The problems of the brain can be complex, but there is an expectation that the situation will improve with the right equipment.

The Big Picture: What's the Importance of This Research?

So, where do we go from here? In the light of the increasing population of the elderly people, unsatisfactory mental health is emerging as a huge public health concern. People do not know whether they have a neurological disease until symptoms show up and even then it is mostly irremediable; however, if brain health can be assessed early and accurately, then it may allow time for intervention. Analogously, consider the difference between catching cancer at its first stage and already stage IV—early detection makes all the difference.

Our formulas have the potential to be of help in screening tests in a clinical setting even more so in resource-constraint environments such as middle-income countries like Pakistan with compromised socioeconomic conditions. Even a few basic tests coupled with standard blood markers could be a home-run in the prevention of brain diseases globally.

In short, this research helped in paving the way toward effective and more personalized brain health care. Whether it's trying to address precursors of Alzheimer's, monitoring changes due to rehabilitative treatments post-stroke, or trying to address barriers and facilitators of cognitive decline in the aged people, these formulations stand the test of time on rampant diseases due to their advantages which are noninvasive and inexpensive. The brain is complicated, but with the right tools, we're starting to crack the code.

Regard Your Brain–Eye as a Combination

Now, how do you expect your brain and eyes to achieve the best performance? In regard to this issue, it goes beyond mere consumption of carrots (though they also have their benefits). Let us find out how one may keep their mind and eyes evergreen and ready to tackle any challenge.

Eye Chart Challenge: Focus on the Eyes for a Change!

The first thing you should do is undergo an eye checkup on a regular basis. Think of it as servicing your car before we have to replace the engine. You wouldn't sit waiting until your car starts smoking before calling a mechanic, would you? Regular checkups in the case of going to the optician can identify problems such as glaucoma, cataracts, or such things as age-related macular degeneration (AMD) very early on and this can help to resolve these in the early disease stage. Unfortunately, eye health is still interlinked with the health of other body systems. For instance, it is widely acknowledged that systemic diseases such as diabetes bring forth diabetic retinopathy, where the tiny blood vessels of the retina are destroyed. Controlled diabetes is associated with a lesser risk of these conditions though. Early detection of such symptoms with regular eye examinations allows one to respond before the vision is permanently destroyed. So, keep your eyes on that chart and make those appointments! Just like your eyes need regular checkups, so does your brain. This can include cognitive screening, memory, or any other function-based testing and even emotional well-being assessments. Folks, be ready to attend these visits, or don't be surprised if there is trouble like vision loss, dementia, or depression.

Brain Games Galore: Work that Brain Like It Is in the Gym!

Want to keep your mind as sharp as your eyesight? Then start treating your brain to a workout. Brain games are equivalent of going for a work out in a gym. More in relation to crossword puzzles and Sudoku; research indicates that engaging in such cognitively challenging activities reduces the risk of dementia among individuals. A neurology research in the year 2019 from the journal *Neurology* distinguishes the fact that the rate of age-related cognitive decay for the individuals who participate in such activities on a regular basis reduces.

In other words, doing even that irritating cross-word puzzle effectively optimizes the operation of the organism's most important organ. Even learning something new— whether it's a new language, how to play an instrument, or mastering that complicated recipe—strengthens your brain's neural connections. It is like a gym for the brain! When the mind is active, even visual processing remains sharp.

Lifestyle Changes: Feed Your Brain

You know the saying, "you are what you eat"? Well, that goes for your brain and eyes too. A diet rich in antioxidants, good fats, fruits, and vegetables improves the heart health and keeps the eyes and the brain active. In fact, certain types of fish such as salmon and tuna contain fatty acids Omega-3 that has been found to help prevent age-related macular degeneration (AMD), as well as dry eyes. Other foods beneficial for the eyes are greens such as zucchinis, spinach, and kale, which are potent suppliers of the lutein and zeaxanthin, two antioxidants that protect your eyes from harmful light.

And of course there is the issue of sleep—your brain's nightly detox session. It has been proven that, during stage "delta," the brain has a system of cleaning the damaging proteins that accumulate in the brain such as beta-amyloid found in an "Alzheimer brain." Yes, sleep is not only a way of increasing your strength levels—you should note, its protective for your brain too. Isn't it a wonder that a person who has had good sleep will have sharper eyesight, because their eyes will be less tense, focused, and ready for whatever task may be at hand.

Lastly, managing stress is crucial. Chronic stress can actually cause visual disturbances like blurred vision or eye strain. If one is to avoid headaches, blurry vision, eye aches, or simply maintaining a more comfortable state of mind, one must set aside some time to relax, de-stress, or take a walk in the garden.

Rest Those Eyes: Second-Sight Trips

Well, let's face it, we are living in the century of computers and video games. Starting with the television and computers and later smartphones, people today have glued their eyes to the screen of electronic devices. Even if it is cool to be watching a great piece of entertainment, screen time can most likely cause eye strain. And it is this moment that one can apply the golden rule of 20-20-20. Namely, every 20 minutes, it is advisable to take 20-second break and focus on an item which is located 20 feet apart. Thus, this technique gives a chance for the muscles of the eye to relax and refocus.

This small effort of closing the eyes can become a big help in preventing the effects of digital eye strain, which is a disorder that affects a significantly large number of people. According to the study by the American Ophthalmological Society conducted at a Medical Institution 58 % of adults said they felt such symptoms of eyestrain after using technology that included headaches, dry eyes, and at worst blurring.

Beyond the Fundamentals: Vision Enhancement for Your Eyes!

Yoga for the eyes: Yes, there is such a thing! If yoga meant stretching and balancing for our muscular anatomy before, it's an understatement. Your eye muscles need a workout too. Simple eye exercises can improve flexibility and reduce strain, keeping your vision sharp. Try doing eye rolls, where you move your eyes in circles without moving your head. Or draw figure eights with your eyes—focus on an imaginary point in the air and trace an eight with your eyes. Another favorite? The near and far focus exercise. Hold your thumb about 10 inches from your face, focus on it, then shift your focus to something farther away, and then back to your thumb. These exercises strengthen your eye muscles, improve focus, and can even help prevent conditions like presbyopia (the age-related decline in near vision). Think of it as yoga for your eyes—a few minutes of practice each day can make a big difference in maintaining clear and healthy vision. But it doesn't stop there—yoga and other physical activities, like walking, running, or dancing, also benefit your brain. They improve blood flow, reduce stress, and increase the production of endorphins, which are the brain's natural mood elevators. Physical activities, including yoga, enhance the connection between body and mind, boost memory, and can even help prevent cognitive decline as you age. They improve blood flow, reduce stress, and increase the production of endorphins, which are the brain's natural mood elevators. Physical activities, including yoga, enhance the connection between body and mind, boost memory, and can even help prevent cognitive decline as you age.

Wrapping It Up: Teamwork Makes the Dream Work

Your brain and eyes are like the dream team of your body. They work hand in hand to help you navigate the world, whether it's reading a book, driving a car, or solving a puzzle. So, it makes sense to keep both in tip-top shape. By following these simple tips—regular checkups, brain-boosting challenges, healthy habits, and some vision-specific exercises—you can keep your brain–eye duo working like a well-oiled machine. After all, you only get one set of peepers and one brain—let's keep them in sync and running smoothly for years to come!

Bibliography

1. Hernandez-Peon R, Sterman MB. Brain functions. Annu. Rev. Psychol. 1966;17(1):363–394. doi: 10.1146/annurev. ps.17.020166.002051.
2. Raichle ME. Two views of brain function. Trends Cogn. Sci. 2010;14(4):180–190. doi: 10.1016/j.tics.2010.01.008.

3. McCaffrey JB. The brain's heterogeneous functional landscape. Philos. Sci. 2015;82(5):1010–1022. doi: 10.1086/683436.

4. Huang L-K, Chao S-P, Hu C-J, Chien L-N, Chiou H-Y, Lo Y-C, et al. Plasma phosphorylated-tau181 is a predictor of post-stroke cognitive impairment: A longitudinal study. Front. Aging Neurosci. 2022;14:1–10. doi: 10.3389/fnagi.2022.889101.

5. Harrabi H, Kergoat M-J, Rousseau J, Boisjoly H, Schmaltz H, Moghadaszadeh S, et al. Age-related eye disease and cognitive function. Invest. Ophthalmol. Vis. Sci. 2015 19;56(2):1217–1221. doi: 10.1167/iovs.14-15370.

6. Subramanian ML, Vig V, Chung J, Fiorello MG, Xia W, Zetterberg H, et al. Neurofilament light chain in the vitreous humor of the eye. Alzheimers Res Ther. 2020;12(1):111. doi: 10.1186/s13195-020-00677-4.

7. Polyakova M, Mueller K, Arelin K, Lampe L, Rodriguez FS, Luck T, et al. Increased serum NSE and S100B indicate neuronal and glial alterations in subjects under 71 years with mild neurocognitive disorder/mild cognitive impairment. Front. Cell Neurosci. 2022;16:1–12. doi: 10.3389/fncel.2022.788150.

Index

Note: Page numbers in *italics* and **bold** refer to figures and tables, respectively.

For Product Safety Concerns and Information please contact our EU
representative GPSR@taylorandfrancis.com
Taylor & Francis Verlag GmbH, Kaufingerstraße 24, 80331 München, Germany

www.ingramcontent.com/pod-product-compliance
Lightning Source LLC
Chambersburg PA
CBHW060555220326
41598CB00024B/3111

* 9 7 8 1 0 3 2 6 4 8 1 6 3 *